IDRC-136e

Role of Cassava in the Etiology of Endemic Goitre and Cretinism

Editors: A.M. Ermans,* N.M. Mbulamoko,† F. Delange,* and R. Ahluwalia‡

*Hopital St-Pierre, Brussels, Belgium
†Institut de recherche scientifique, Kinshasa, Zaire
‡Program Officer, Health Sciences Division, IDRC, Ottawa, Canada

Contents

Foreword

During the last 10 years, intensive efforts and investments have been made on an international scale into research to improve the protein content and yield of cassava (*Manihot esculenta* Crantz), to combat the diseases to which the cassava plant is susceptible, and to improve the processing methods.

Cassava plays a dominant role in feeding low-income people and in many tropical countries forms the basic ingredient of their diet. It is well known that cassava contains linamarin, a cyanogenic glucoside, which, when acted upon by the enzyme linamarase contained in the plant, forms cyanide, which in humans is predominantly converted to thiocyanate, a known goitrogenic agent. This cycle of events can be prevented by adequate processing of the cassava root so that the hydrocyanic acid is released from the plant. In rural areas, different processing methods have been used over time with various results. In recent years, in parts of Zaire where food scarcity exists it has been observed that fewer people are using even traditional methods of processing. Instead; the cassava root is being consumed immediately after it has been extracted from the soil. This results in high doses of thiocyanate being found in humans and is associated with chronic conditions of the thyroid and central nervous system.

The data discussed in the monograph confirm that the prevalence of goitre is related to the balance between thiocyanate and iodine intake. Endemic goitre and related cretinism is an indication of a more serious public health problem, especially among pregnant women, because a high percentage of their offspring suffer from congenital hypothyroidism (15%).

It is, therefore, essential that greater emphasis be placed on the development of new varieties of cassava that contain minimal amounts of linamarin.

The research carried out, under extremely difficult field conditions, by the Zairians and Belgians has been of very high quality. Further investigations are under way, and a supplementary report on the final phase of this research is expected to be published in 1981.

J. Gill, MD
Director
Health Sciences Division
IDRC

Préface

Malgré son passé relativement récent, la coopération technique et scientifique entre le CEMUBAC, le CRDI et l'IRS possède déjà une histoire riche en événements de divers ordres.

En effet, cette coopération n'a cessé de s'élargir et de s'approfondir, depuis la convention signée à Lwiro le 6 décembre 1972 entre l'ex-IRSAC et le CEMUBAC. Deux ans plus tard, soit le 23 janvier 1974, une convention complémentaire, signée également à Lwiro, était consacrée plus particulièrement au "Programme National d'Éradication du Goitre et du Crétinisme Endémiques au Zaïre". Les deux conventions ont trouvé une de leurs premières consécrations par l'organisation, du 23 au 28 septembre 1974, du Colloque de Gemena consacré à la "Stratégie intégrée d'éradication du goitre et du crétinisme endémiques au Nord Zaïre", colloque dont l'un des animateurs principaux était le Professeur Ermans, à qui nous rendons un hommage mérité.

Depuis lors, des contacts fructueux se sont poursuivis, et aujourd'hui, on peut affirmer que l'un des résultats de cette coopération, à laquelle est étroitement lié le CRDI, se trouve être la présente monographie, consacrée à l'ensemble des activités scientifiques du programme IRS-CEMUBAC en général, et aux recherches sur l'action toxique du manioc en particulier.

Nul n'ignore que le manioc sert d'aliment de base dans de nombreuses régions tropicales et subtropicales. Son influence, dans le cadre de la nutrition, sur la santé des populations de ces contrées, est une des préoccupations majeures des chercheurs des différents pays concernés.

Au Zaïre, dans le cadre de la programmation de la recherche, l'IRS a inscrit parmi ses problèmes prioritaires, les recherches dans le domaine de la santé, qui prévoient des études pour l'amélioration de la situation nutritionnelle de la population.

Ces préoccupations — il faut le préciser — font partie d'un ensemble de questions relevant de la politique scientifique en général et de la mission spécifique de l'IRS en particulier.

Sans entrer dans les détails, il nous suffit de rappeler que notre Institut, ayant sa vocation spécifique de recherche définie par les différents textes de loi le régissant, a été chargé par ailleurs d'une haute mission dans le cadre du plan national de développement et de l'exécution de la politique scientifique.

Tout responsable zaïrois — comme tout responsable de n'importe quel pays organisé — est conscient de l'importance de la science et de la technologie dans le développement national, à la fois comme moteur et stimulant "novateur dans la production de biens et de services", mais aussi, comme le signale un rapport de l'UNESCO, comme "méthode d'attaque d'une foule de problèmes qui se posent aux gouvernements dans les domaines les plus divers : politique, social, économique, militaire, etc. . .". A ces domaines il faut ajouter naturellement celui de la santé et de la médecine, objet propre et particulier de la coopération entre le CEMUBAC et l'IRS.

Dans le cadre de l'exécution de la politique scientifique nationale et de la réalisation de son programme d'action, l'IRS bénéficie du concours d'un certain nombre de pays, dont la Belgique, le Canada, la France, le Japon, la République populaire de

Chine, la Yougoslavie, les États-Unis d'Amérique, etc. Parmi les organisations internationales, l'UNESCO mérite une place de choix. Enfin, sur le plan des échanges d'informations scientifiques, il existe un contact plus ou moins régulier entre l'IRS et un certain nombre d'organismes semblables fonctionnant dans d'autres pays d'Afrique, d'Europe, d'Asie, d'Amérique et d'Océanie.

En ce qui concerne plus précisément l'objet de la présente monographie, il résulte de la collaboration entre le CEMUBAC, le CRDI et l'IRS. Celle-ci a déjà permis l'organisation de plusieurs rencontres scientifiques, la réalisation de certains projets mixtes et la publication de quelques résultats, notamment dans le domaine de la malnutrition et du goitre endémique.

Les chapitres de ce volume constituent le fruit du travail d'une équipe de chercheurs zaïrois et étrangers.

L'étude a été réalisée dans un esprit où les besoins médicaux et sanitaires des populations concernées et les impératifs de la recherche scientifique ont été soigneusement balancés.

Elle débouche sur des conclusions scientifiques dont la portée dépasse les frontières du Zaïre, puisqu'elles concernent le rôle joué par le manioc dans le déclenchement des maladies thyroïdiennes et que le manioc est largement consommé dans de nombreuses régions tropicales et subtropicales. C'est donc un travail qui aura un impact considérable sur le plan sanitaire, médical, social et économique non seulement au Zaïre, mais aussi dans des pays, qui comme le nôtre, intègrent l'action de la recherche scientifique et technologique dans le processus du développement socio-économique et socio-culturel.

N.M. MBULAMOKO
Président du
Conseil scientifique de
l'Institut de recherche scientifique

Contributors

P.A. Bastenie, MD, PhD, Dept of Experimental Medicine, University of Brussels, Belgium.

H. Berquist, MD, Hospital of Karawa, Evangelic Community of Ubangi–Mongala, Zaire.

P. Bourdoux, PhD, Senior Chemist, Dept of Radioisotopes, St. Pierre Hospital, University of Brussels, Belgium; IDRC Consultant.

P. Courtois, MD, CEMUBAC–AGCD, Centre of Gemena, Zaire.

N. Cremer, MD, Head, Dept of Pediatric Radiology, St. Pierre Hospital, University of Brussels, Belgium.

F. Delange, MD, PhD, departments of pediatrics and radioisotopes, St. Pierre Hospital, University of Brussels, Belgium; Codirector of the Goitre Program IRS–CEMUBAC, Zaire.

K.D. Döhler, MD, Dept of Clinical Endocrinology, University of Hannover, Germany.

M. Dramaix, Dept of Epidemiology and Social Medicine, University of Brussels, Belgium.

A.M. Ermans, MD, PhD, Head, Dept of Radioisotopes, St. Pierre Hospital, University of Brussels, Belgium; Director, Goitre Program IRS–CEMUBAC, Zaire.

S. Filetti, MD, Institute of Endocrinology, University of Catania, Italy.

M. Gérard, MD, CEMUBAC–AGCD, Centre of Gemena, Zaire.

A. Hanson, Chemist CEMUBAC–IDRC, Centre of Gemena, Zaire.

R.D. Hesch, MD, PhD, Dept of Clinical Endocrinology, University of Hannover, Germany.

J. Kinthaert, Dept of Radioisotopes, St. Pierre Hospital, University of Brussels, Belgium.

R. Lagasse, MD, CEMUBAC–AGCD, IDRC Consultant, Dept of Epidemiology and Social Medicine, University of Brussels, Belgium.

K. Luvivila, MD, Dept of Public Health, IRS, Zaire; Codirector of the Goitre Program IRS–CEMUBAC, Centre of Gemena, Zaire.

M. Mafuta, Chemist, IRS, Centre of Gemena, Zaire.

G. Nelson, Nurse, Hospital of Karawa, Evangelic Community of Ubangi–Mongala, Zaire.

V. Pezzino, Institute of Endocrinology, University of Catania, Italy.

L. Ramioul, Dept of Epidemiology and Social Medicine, University of Brussels, Belgium.

G. Roger, Psychologist, CEMUBAC; IDRC–AGCD Consultant, University of Brussels, Belgium.

P. Seghers, Chemist, CEMUBAC–IDRC, Centre of Gemena, Zaire.

E. Simons-Gérard, Economist, CEMUBAC, University of Brussels, Belgium; IDRC Consultant.

S. Squatrito, Institute of Endocrinology, University of Catania, Italy.

C.H. Thilly, MD, Dept of Epidemiology and Social Medicine, University of Brussels, Belgium; Codirector of the Goitre Program IRS–CEMUBAC, Zaire.

F. Trimarchi, Institute of Medical Pathology, University of Messina, Italy.

D. Tshibangu, MD, AGCD Consultant, Goitre Program IRS–CEMUBAC, Zaire.

L. Vanderlinden, Chemist, Dept of Radioisotopes, St. Pierre Hospital, University of Brussels, Belgium.

J.B. Vanderpas, MD, CEMUBAC–AGCD, Centre of Gemena, Zaire.

M. Van der Velden, Dept of Radioisotopes, St. Pierre Hospital, University of Brussels, Belgium.

N. van Minh, MD, Research Fellow, Dept of Radioisotopes, St. Pierre Hospital, University of Brussels, Belgium.

R. Vigneri, MD, Institute of Endocrinology, University of Catania, Italy.

Y. Yunga, MD, IRS Codirector of the Goitre Program IRS–CEMUBAC, Centre of Gemena, Zaire.

Acknowledgments

The interdisciplinary study described in this monograph was largely carried out in two remote areas of Zaire, where we were called upon to solve numerous logistic, technical, administrative, and human problems in addition to medical and scientific investigations. This work was only completed because of the generosity and help of a series of organizations to whom we would like to express our deep appreciation: the International Development Research Centre (IDRC, Ottawa, Canada), l'Institut de la Recherche Scientifique (IRS) and le Commissariat d'État à la Santé Publique of the Republic of Zaire, l'Administration Générale de la Coopération au Développement (AGCD, Belgium), le Centre Scientifique et Médical de l'Université Libre de Bruxelles pour ses Activités de Coopération (CEMUBAC), OXFAM, the American Peace Corps, and, finally, within the University of Brussels, the departments of radioisotopes and pediatrics (Hôpital St. Pierre) and the School of Public Health.

The study of the toxicity of cassava was essentially supported by the IDRC whose funding made possible the establishment of a technologically sophisticated research unit in the heart of Ubangi and the creation of the necessary scientific infrastructure. We are particularly grateful to Dr J. Gill and R. Ahluwalia, whose interest in our work, helpful criticism, continuous administrative aid, and friendship were decisive in ensuring the success of our project. We would also like to thank the other divisions of the IDRC that collaborated on the project, especially the Communications Division, its director, R. MacIntyre, and one of the technical editors, A. Chouinard, for the preparation and publication of this monograph.

The realization of the program was the result of fruitful collaboration between our university and l'Institut de la Recherche Scientifique in Zaire. The IRS set in motion a shift of our activities from Kivu to Ubangi, the centre of the most severe goitre endemia in Zaire. The IRS made possible the creation of the principal operational base in Gemena as well as a research laboratory. For this initiative, we thank Dr Ntika-Nkumu, former Délégué Général of the IRS. His successors, Citoyen Kama Funzi Mudindambi and Professor Iteke Fefe Bochoa, the present Délégué Général of the IRS continued this collaboration with enormous interest. We would also like to acknowledge the devoted help of the technical and administrative personnel of the IRS.

In the context of the campaign against endemic goitre, we have received support from the Commissariat d'État à la Santé Publique, and, in particular, from its Secrétaire d'État, Dr Lekie Bottee. We extend our thanks to him and to doctors Luvivila and Nzila, the former and present médecins sous-régionaux in Gemena. The nurses of the Centre of Gemena and those working in the rural dispensaries have provided very capable help.

The treatment program in Ubangi was made possible by the financial aid of the AGCD. To all the Belgian officials who sustained this effort goes our gratitude, particularly to doctors M. Kivits, J. Burke, and J. Mahaux. We would also like to thank the Fonds de Médecine Tropicale (FOMETRO) for their support, especially Professor S. Halter, President, and Dr J.F. Ruppol, R. Ramet, and J. Allen in Kinshasa.

Work in the field benefited from the generous and very valuable help of the Evangelic Community of Ubangi–Mongala and, in particular, the staff of the Hospital of Karawa. We found the same welcome and support in the "Paroisse de Bominenge."

We are grateful to the American Peace

Corps in Zaire, its directors of public health, M. Robbins and, subsequently, T. Manchester, and to the volunteers who worked with our team in Gemena, O. Levesque, J. Marquis, S. and B. Overman, M. Kocurek, C. Glahn, and R. Coismain.

The program received support from a number of other sources that were a great help in the different stages of work, notably OXFAM, the Fonds de la Recherche Scientifique Médicale (FRSM, Belgium, contracts 3.4565.75 and 3.4553.75), the Banque Nationale de Belgique (Belgium), the World Health Organization (WHO, Geneva, contract R00162), the Institut Belge de l'Alimentation et de Nutrition (IBAN), the International Atomic Energy Agency (IAEA, Vienna, contract 216/R1/0B), and the Ministère de la Politique Scientifique (Belgium) within the framework of the Association Euratom–University of Brussels–University of Pisa.

The basic structure in our university supporting this work is the CEMUBAC. The President, Professor M. Millet, and the General Secretary, Professor R.E. De Smet, were instrumental in initiating the project. We also benefited from the generous aid of Professor H.L. Vis, notably for all the work carried out in Kivu, and Dr Matundu-Nzita, Director of CEMUBAC Zaire, as well as doctors J. Massart, J. Szeles, and R. Sannon, CEMUBAC representatives in Kinshasa.

The academic authorities of the Université Libre de Bruxelles have encouraged us throughout. We wish to thank professors P. Foriers and J. Michot, Recteurs, and most especially Professor A. Jaumotte, Président du Conseil d'Administration. He testified to his interest in our project by participating in the inauguration of the IRS–CEMUBAC centre in Gemena. We are most grateful for his support, manifested also in his willingness to accept the presidency of CEMUBAC in Belgium. We appreciate the sustained interest of several members of the Faculty of Medicine, professors A. Dachy, A. Loeb, and H.L. Vis, chiefs of the Department of Pediatrics, Professor M. Graffar, Director of the Department of Epidemiology and Social Medicine, Professor J.E. Dumont, Director of the Institut de Recherche Interdisciplinaire en Biologie Humaine et Nucléaire (IRIBHN).

We would like to thank all our close collaborators, both in the Department of Radioisotopes and in the School of Public Health. In many capacities they contributed to the realization of an enormous number of chemical determinations and their tabulation. We particularly thank C. Branders, E. Colinet, M.C. Deliège, B. de Poortere, D. Dewilde, M. Fernandez, H. Gheyssens, N. Soupart, and G.J. Vincke. We appreciate greatly the aid of our secretaries, E. Andries, O. Dedycker, and G. Vandenbosch; their efficiency made possible the resolution of any number of administrative and logistic problems posed during the course of the program.

We sincerely thank Dr E. Schell-Frederick for her skillful translation of this monograph. Finally, we are deeply indebted to Professor G. Podoba and his research group in Bratislava, Czechoslovakia, who in 1967 focused our interest on the action of naturally occurring goitrogens.

Some tables and figures were reproduced with the authorization of the *Journal of Clinical Endocrinology and Metabolism, American Journal of Clinical Nutrition*, and the *Journal of Endocrinological Investigation.*

Introduction

A.M. Ermans

This monograph summarizes, partially in chronological order, a series of clinical and experimental studies of endemic goitre and cretinism. These studies have established that goitre and cretinism in Central Africa are caused by the joint effects of iodine deficiency and a goitrogen found in cassava.

The orientation of these studies was in part determined by fortuitous circumstances. The initial aim was to evaluate a new procedure for iodine supplementation adapted to the rural populations of Zaire. During preliminary studies preceding the institution of this prophylactic program, we observed an important discrepancy between the prevalence of goitre on the one hand and the biologic parameters defining the severity of iodine deficiency on the other. The intervention of another etiologic factor, distinct from iodine deficiency, was probable. Its identification was systematically sought. Nutrition surveys and a series of clinical investigations focused our attention on the possible role of cassava. Specifically, we observed that cassava ingestion in humans induces definite abnormalities in thyroid function, qualitatively identical to those resulting from the administration of thiocyanate. This hypothesis seemed particularly attractive in view of the then recent findings of Ekpechi (1967) demonstrating in the rat a definite goitrogenic effect of prolonged cassava administration. Available data indicated that cassava does not contain thiocyanate or its precursors. In contrast, it does contain a cyanogenic glucoside, linamarin, sometimes in large quantities, whose degradation releases hydrocyanic acid. The conversion of this highly toxic substance in the human body leads to the formation of thiocyanate.

The hypothesis that cassava plays a causative role in endemic goitre and cretinism was based on well-documented experimental findings. However, the crucial question was whether cassava actually intervenes under the conditions found in Central Africa populations or whether it is merely a fortuitous coincidence.

The intervention of nutritional goitrogens in human pathologic conditions has often been suggested but has never been convincingly demonstrated. Although the opinion prevails that normally humans do not ingest sufficient amounts of these substances to induce a significant antithyroid effect, cassava is a staple food for many populations and may be implicated in endemic goitre. Several objections have been raised about the hypothesis of a goitrogenic action of cassava in humans.

In the first place, the quantity of thiocyanate that is goitrogenic in humans is clearly greater than that resulting from cassava ingestion. The serum thiocyanate concentration in the goitrous populations of Ubangi and Kivu attains a mean value of 10 μg/ml. No clinical or experimental studies have shown an inhibitory effect on the thyroid iodide pump at such low concentrations. In addition, in these populations, thyroid radioactive iodine uptake is extremely elevated, thus rendering improbable the hypothesis that the iodide pump would be influenced by moderate increases in serum thiocyanate concentration. Second, the precise extent of iodine deficiency is difficult to estimate. One cannot exclude the possibility that local differences in iodine supply explain the observed epidemiological variations and have not been detected by current methods. Moreover, the estimation of the respective roles played by iodine deficiency and thiocyanate overload is complicated because these two factors produce similar changes in thyroid function under chronic conditions.

Third, in many areas where cassava is widely consumed, there is no abnormal prevalence of goitre. The toxicity of cassava is well recognized by African populations and they employ different detoxification procedures during preparation of the cassava roots. In principle, these procedures ensure complete release of the hydrocyanic acid. Finally, another possible contributing factor is the consumption of other plants containing thioglucosides, the direct precursors of thiocyanate. It has been impossible to exclude, a priori, that these foodstuffs rather than cassava play a predominant role in the observed thiocyanate overload.

The observations presented in this monograph have ruled out most of these objections unequivocally. We have demonstrated in humans and in rats that thiocyanate formed from the catabolism of linamarin disrupts the mechanisms by which the thyroid adapts to iodine deficiency and aggravates the effect of iodine deficiency on thyroid function. In addition, we have elucidated the nature of the inhibition of the iodide pump produced by these low thiocyanate concentrations and have explained previous failures to detect the inhibition.

The investigations reported in this monograph were in fact carried out in successive stages on the basis of an alternation between clinical and experimental studies; the experimental results obtained enabled us to define optimally the procedures of clinical investigation. However, with a view to simplifying the presentation of our results, we have reported clinical and experimental studies in distinct chapters.

One of the major problems confronted in this study was endemic cretinism, a true scourge in these populations because of its high incidence and marked severity. Its pathogenesis is directly linked to alterations in thyroid function in the fetus and newborn but has not yet been clearly defined. We were able to study pregnant women and newborns in the maternity section of the hospital of Karawa in Ubangi. The following questions were addressed: What is the activity of the thyroid in the newborn at the time of birth? What evidence of thiocyanate overload is present in the fetus and in the nursing infant? How important are the abnormalities in thyroid function in the mother? And what are the effects of the goitrogenic environment on development in utero, perinatal survival, and subsequent psychomotor development? Clinical observations were supplemented by an experimental study on the effects of goitrogens in rats at birth and during lactation.

One of the major implications of our study is the necessity to reduce the hydrocyanic acid concentration in the cassava consumed by these populations. This objective could be reached either by modifying the methods of preparing the food or promoting the cultivation of cassava with lower linamarin concentrations. These are long-term solutions. Because of the severity of the epidemiological situation, it seemed imperative to initiate a short-term remedy — iodine supplementation. Correction of the iodine deficiency completely antagonized the antithyroid effects resulting from cassava ingestion. A pilot study was first carried out on Idjwi Island in Kivu to establish an optimal treatment protocol. In a second stage, iodine prophylaxis was extended to more than half a million inhabitants of the equatorial region. The programing and evaluation of such a campaign in remote regions posed a series of difficult organizational and logistic problems, as iodized oil had never before been administered on such a large scale. This particular public health aspect of our activity is described in a separate chapter.

Chapter 1

Cassava Toxicity: the Role of Linamarin

P. BOURDOUX, M. MAFUTA, A. HANSON, AND A.M. ERMANS

Cassava is a shrubby tree of the euphorbiaceous family. The plant is of American origin and was introduced into Africa by the slave-traders nearly 3 centuries ago. *Manihot esculenta* Crantz (*M. utilissima*) is the most common variety. This plant, essentially tropical, spread rapidly throughout the sub-Sahelian belt. It currently is distributed almost symmetrically on either side of the equator. Cassava requires soil that is rich in humus. Thus, the primary forest, once cleared, is an ideal environment for this plant. Unfortunately, cassava cultivation rapidly depletes the soil and contributes to the destruction of the forests. Cassava has achieved considerable agricultural importance as the major source of tapioca and fodder for cattle, particularly in the European Economic Community (Nestel 1973; Phillips 1974). It is almost exclusively cultivated in developing countries and represents the essential source of calories for more than 300 million people living in the tropics (Nestel 1973).

Zaire is currently ranked as the third largest producer of cassava (Nestel 1973), the roots of which are an important source of food throughout the country. In regions such as Ubangi (northwest Zaire), where cassava is consumed as a staple food, certain nutritional problems have been noted, specifically endemic goitre and cretinism (chapter 6).

Toxicity, Cyanogenesis, and Cyanide

The autochthonous populations recognized long ago the toxic properties of certain plants. Shortly after the introduction of cassava, they began to consider its toxicity to be parallel with the bitterness of the roots. The characteristic odour of bitter almonds indicates the origin of the toxicity, which can be attributed to the liberation of hydrocyanic acid (HCN).

Other plants are also known to give off HCN. In their chemical taxonomy, Dilleman (1958) and Hegnauer (1963) mention approximately a thousand plants exhibiting this property. However, actual synthesis and production of HCN have never been demonstrated in such plants (Dilleman 1958; Conn 1978). The cyanide-forming plants contain substances capable of liberating HCN under certain conditions. The hydrolysis of these substances produces one or more nonhydrolyzable reducing sugars (most often glucose) and a nonglucidic moiety, an aglycone. The bond between the glucide and the aglycone can be located on an asymmetric carbon, giving rise to alpha- and beta-heterosides. Because of their chemical composition, these substances are called cyanogenic glucosides. The majority of the cyanogenic glucosides are beta-glucosides.

Several plants, used as food, contain cyanogenic glucosides — beans, corn, sorghum, sweet potatoes, lettuce, peas. With the exception of cassava and certain beans (*Phaseolus lunatus*), these glucosides are localized either in the inedible portions of the plant or in the edible portions but in such small quantities that they are fit for consumption.

The reason for the presence of toxic products in plants is a controversial point. It could be the accumulation of products of the

Linamarin

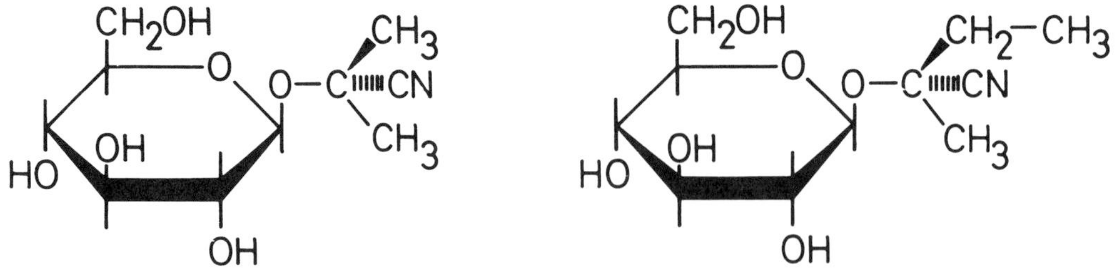

melting point: 143–144 °C (uncorrected)
$[\alpha]_D^{32}$: −28.5 (C = 3.86; H_2O)
crystallization: ethyl acetate
molecular weight: 247.2

Lotaustralin

melting point: 123.5–124.5 °C
$[\alpha]_D^{25}$: −19.15 (C = 1.07; H_2O)
crystallization: ethanol-hexane
molecular weight: 261.2

Fig. 1. *Structure and physicochemical properties of the cyanogenic glucosides (linamarin and lotaustralin) from cassava.*

catabolism of amino acids (Conn 1973b) or a mechanism of defence for the plant against predators. The increased content of cyanogenic glucosides in old plants corresponds most frequently to an increased nitrogen metabolism. The catabolism of cyanogenic glucosides could represent an alternative to the nitrogen cycle.

The existence of cyanogenic glucosides has also been demonstrated in the animal kingdom, for example, Myriapoda and Zygaeninae (Eyjolfsson 1970), serving simultaneously as a mechanism of defence and of attack.

The habitual distinction between the bitter and sweet varieties of cassava is based on the cyanide content of the edible portions of the roots. Bolhuis (1954) arbitrarily classified roots as nontoxic or sweet (<50 mg HCN/kg fresh weight), intermediate (50–100 mg HCN/kg fresh weight), and toxic or bitter (>100 mg HCN/kg). In reality, the phenomenon is more complex, and there is no direct relation between the bitterness of the roots and their content of cyanogenic glucosides (de Bruijn 1971). Our experience indicates no sure or rigorous distinction based on the morphological characteristics of the plant. This conclusion agrees with Bolhuis (1954) and Coursey (1973).

In the rural sector of Ubangi, the inhabitants distinguish between bitter and sweet roots in an intuitive manner, and they are capable of determining the more toxic of two plants. The problem is that a "bitter" root in an area where the overall crop has relatively low cyanide concentrations may be considered as "sweet" in another area where concentrations are generally higher. For example, in Bavula, cassava roots distinguished as bitter and sweet respectively contained 44.1 and 12.5 mg HCN/kg fresh weight, whereas, in Lebo, the bitter roots had 160.1 and the sweet 20.1 mg HCN/kg fresh weight. Moreover, the cyanogenic glucoside content (and thus the HCN content) of cassava varies widely as a result of environmental factors: season, type of soil, presence of certain ions, diurnal fluctuations, genetic factors, etc. (de Bruijn 1971).

Linamarin

In 1906, Dunstan et al. isolated a cyanogenic glucoside from cassava roots, identifying it as the same glucoside extracted from flax (*Linum usitatissimum*) (Jorissen and Hairs 1891) and from a variety of beans (*Phaseolus limensis*) (Dunstan and Henry 1903). For this reason, it was called linamarin or linamaroside. In cassava, there is also a small amount of another glucoside, lotaustralin, also isolated from *Lotus australis* (Finnemore et al. 1938). The ratio of linamarin to lotaustralin in cassava is reported to vary from 93:7 to 97:3 (Butler 1965; Jansz et al. 1974a). The chemical structure of these two glucosides, although simple, was not defini-

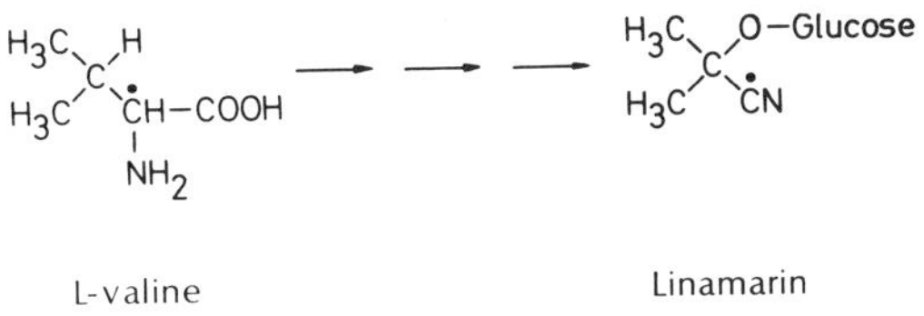

L-valine Linamarin

Fig. 2. Similarity between the structure of L-valine and linamarin.

tively established until the 1960s (Clapp et al. 1966; Bisset et al. 1969). The structure and physicochemical properties of these two substances are shown in Fig. 1.

Biosynthesis of cyanogenic glucosides

In the last 2 decades a number of studies on the biosynthesis of cyanogenic glucosides have been carried out, principally by Conn (1973a). The utilization of labeled precursors has contributed greatly to an understanding of the pathways involved. It was shown that dhurrin is formed from tyrosine (Conn and Akazawa 1958). Subsequently the incorporation of other amino acids whose structures resemble the aglycone of cyanogenic glucosides was also demonstrated (Mentzer et al. 1963; Ben Yehoshua and Conn 1964; Abrol and Conn 1966; Abrol 1967; Nartey 1968). Nartey (1968) demonstrated that L-valine could serve as a precursor for linamarin (Fig. 2). Some intermediate steps in the biosynthesis of cyanogenic glucosides are still uncertain. According to Conn and Butler (1969), the pathway shown in Fig. 3 is consistent with experimental results.

Liberation of hydrocyanic acid

It is well established that plant tissues containing a cyanogenic glucoside also contain one or more enzymes for its decomposition. However, some exceptions to this generalization have been reported (Finnemore and Gledhill 1928; Cooper-Driver and Swain 1976; Secor et al. 1976). In cassava, HCN is most easily liberated by the enzyme linamarase or linase, also contained in the plant. Contact between the substrate (linamarin) and the enzyme is sufficient to initiate hydrolysis of the glucoside and liberation of HCN. Contact is made by simple wounding of the tissues or by any process that ruptures the cell walls (grinding, freezing, adding chemical agents). The pathway of linamarin degradation is shown in Fig. 4.

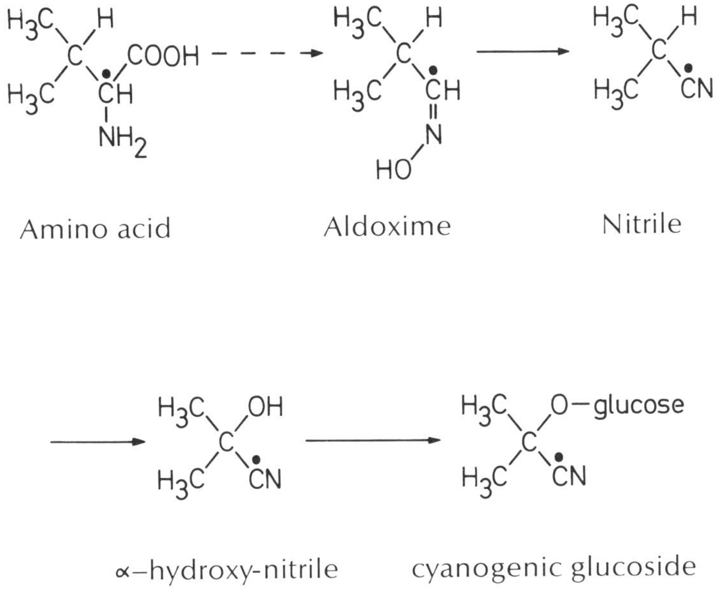

Amino acid Aldoxime Nitrile

α-hydroxy-nitrile cyanogenic glucoside

Fig. 3. Biosynthesis of a cyanogenic glucoside (linamarin) from an amino acid (L-valine) precursor according to Conn and Butler (1969).

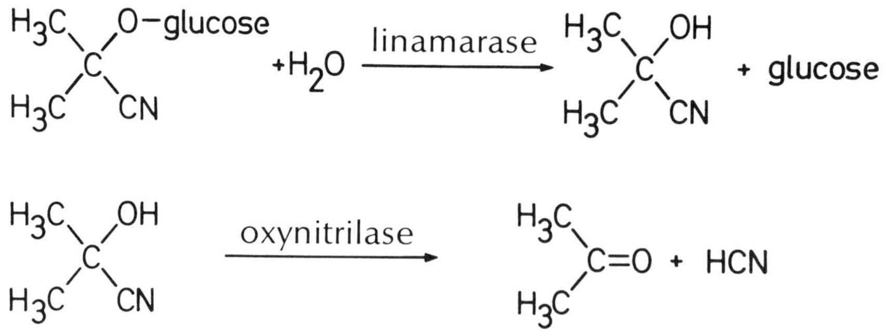

$$(H_3C)_2C(\text{O-glucose})(CN) + H_2O \xrightarrow{\text{linamarase}} (H_3C)_2C(OH)(CN) + \text{glucose}$$

$$(H_3C)_2C(OH)(CN) \xrightarrow{\text{oxynitrilase}} (H_3C)_2C{=}O + HCN$$

Fig. 4. Enzymatic degradation of linamarin according to Conn (1969).

Linamarase

Linamarase belongs to the group of beta-glucosidases. The enzyme found in cassava is extremely specific (Butler et al. 1965). The other natural beta-glucosidases, e.g., emulsin, provoke little or no hydrolysis of linamarin. Linamarase is also present in other plants, i.e., *Linum usitatissimum, Trifolium repens*, etc.

Very recently, a method was described for extensive purification of linamarase extracted from cassava (Cooke et al. 1978). The method, based on DEAE-cellulose chromatography, permits a 350-fold increase in specific activity with reasonable recovery of activity (35%); de Bruijn (1971) has reported that the amount of enzyme present in cassava is sufficient to account for a nearly total hydrolysis of linamarin. According to this work, the addition of exogenous enzyme does not influence, or only slightly influences, the amount of HCN liberated. In contrast, Jansz and Nethsingha (1978) have concluded that the low enzyme content is not sufficient for complete hydrolysis. The preparation of linamarase of high specific activity should resolve this question. It should be noted that Joachim and Pandittesekere (1944) mentioned that linamarase decomposes at 72 °C.

Determining cyanogenic glucoside content

The detection of cyanogenic glucoside in plant tissues is almost universally done by indirect means, i.e., by detection of the amount of HCN liberated after hydrolysis. The method we used was based on the establishment of contact between the substrate, linamarin, and the enzyme, linamarase, both of which are present in cassava. Contact was established by maceration. The process was essentially one of autolysis.

Method: The whole cassava root (or other plant sample) was peeled and rapidly homogenized. An aliquot of the homogenate was weighed and transferred to a 25 ml Erlenmeyer flask with a centre-well containing 0.5 ml NaOH 1 M. Five millilitres acetate buffer 0.1 M, pH 5.5 were added; the flask was closed hermetically and incubated at 37 °C for 20 hours with agitation. The HCN liberated by autolysis was trapped in the centre-well and measured in an aliquot.

Two methods have been developed for the determination of cyanide content:

- Direct potentiometric method: in recent years, the development of specific electrodes has made possible the quantitation of certain ions. We used a cyanide-specific electrode (Orion 94-06A), diluting the contents of the centre-well, after autolysis, with KOH 0.1 M to obtain a sufficient volume. The potential difference observed was compared with a standard curve established by dissolving known quantities of cyanide in KOH 0.1 M.
- Spectrophotometry: the procedure of Aldridge (1945) was followed.

The two methods give comparable results ($r=0.997$; $P<0.001$; $n=33$), but for practical reasons the second has been the method of choice. Recently a method allowing the semi-quantitative estimation of linamarin itself in plant or animal tissues has been proposed (Zitnak et al. 1977).

Conditions of autolysis

Autolysis brings the enzyme, present in the plant, in contact with its substrate. Thus, the conditions of autolysis must reveal full activity of the enzyme to produce maximal glucoside hydrolysis.

The optimum pH for linamarase is between 5.0 and 6.0 (Seifert 1955), thus dictating use of an acetate buffer (0.1 M, pH 5.5). However, there is limited enzyme activity in the absence of a buffered medium: in distilled water, bitter and sweet roots (12 determinations each) liberated, respectively, 36 ± 12 (mean ± SD) and 5 ± 1 mg HCN/kg fresh weight, whereas in acetate buffer the figures were 142 ± 6 and 27 ± 6 mg HCN/kg fresh weight. This observation contradicts the findings of de Bruijn (1971).

Losses during homogenization

The liberation of HCN, in principle, results uniquely from the contact between the substrate and enzyme. Thus it is important to determine the amount of HCN lost between homogenization and the beginning of the incubation period. In our study, incubation was initiated at various intervals following homogenization (1–270 minutes, each time being measured in quintuplet) in aliquots of the same homogenate. The results are presented in Table 1. Unexpectedly, we found that the HCN content was little affected when incubation was initiated within 60 minutes of homogenization.

Retention of cyanide ion in the incubation medium was minimized by the use of small quantities of material. However, Table 2 illustrates that this source of error is not negligible. This experiment contradicts the work of Pieris et al. (1974) who reported that even important modifications in the amount of material did not influence cyanide liberation, expressed per gram of sample. This discrepancy could be explained by the presence of an excess of exogenous linamarase in the experiments of Pieris.

Table 1. Amount of HCN liberated after different intervals between homogenization of the root and incubation at 37 °C.

Interval (min)	HCN liberated during incubation: mg/kg fresh weight (mean ± SD)	%
1	176 ± 20	100
30	174 ± 30	99
60	174 ± 23	99
90	138 ± 38	78
270	62 ± 13	35

Table 2. Influence of weight of the cassava sample on HCN content (mean ± SD).

Sample (g)	HCN (mg/kg)
0.1	16 ± 2
0.5	11 ± 1
1.0	13 ± 1
1.5	12 ± 1
2.0	11 ± 1

Cyanide content

Several authors (de Bruijn 1971 and cited references) have shown that the glucoside content varies between the proximal (near the peduncle) and the distal parts of the roots, although the variations appear to be purely random.

We divided each of 21 roots from different plants into eight equal parts. Each part was homogenized and the cyanide content measured in triplicate (Table 3). Samples 1–12 were taken from plants considered by the natives to be bitter, samples 13–21 from plants considered to be sweet.

In some roots the proximal part was richer in cyanide. In others, the inverse was found, and in others the cyanide content was constant or completely irregular. Although, on average, the cyanide concentrations were higher in bitter than in sweet plants, there was an important overlap between the two samples, confirming the absence of a correlation between the bitterness of cassava and its content of hydrocyanic acid (de Bruijn 1971 and cited references).

We also determined the cyanide content in all the roots of two different plants from the experimental station of Boketa (Ubangi)

Table 3. Variations in cyanide content (mg HCN/kg fresh weight) from proximal end (near peduncle) to distal end of 21 peeled roots from Ubangi.

Root Number	Distal (8)	7	6	5	4	3	2	Proximal (1)
1	41.5	75.6	77.9	86.6	81.1	93.1	127.2	143.7
2	48.0	45.5	45.1	43.3	44.4	31.5	45.9	37.2
3	27.2	16.2	13.8	11.7	8.3	10.4	15.9	8.1
4	37.0	50.0	53.0	66.5	69.5	84.2	134.2	206.7
5	111.7	105.8	24.4	34.0	83.9	87.9	89.7	108.3
6	80.4	107.2	105.8	107.7	103.3	127.0	137.9	178.3
7	37.1	50.1	53.0	66.5	69.6	84.3	134.3	206.8
8	82.8	84.9	79.0	67.4	73.1	88.5	83.2	72.7
9	137.1	100.2	75.4	36.5	38.2	41.2	64.5	35.4
10	64.0	113.2	196.5	169.5	183.9	101.5	85.6	62.8
11	176.2	145.3	195.9	189.1	162.1	190.5	202.3	174.1
12	85.9	84.1	78.4	73.3	70.7	69.2	61.5	74.5
13	41.5	45.2	48.4	55.0	60.0	73.8	74.5	84.0
14	53.7	51.7	53.4	61.6	48.5	49.7	60.3	—
15	59.9	66.9	85.2	65.1	75.2	75.5	104.9	87.9
16	9.7	7.8	6.4	4.9	2.5	3.9	5.1	7.6
17	18.0	34.7	23.8	25.7	21.0	26.8	31.0	48.3
18	57.5	49.3	54.4	55.1	53.1	64.1	53.5	43.2
19	8.3	13.4	11.7	20.4	18.7	24.3	27.3	6.3
20	66.4	61.2	65.2	65.4	54.9	73.1	78.9	78.9
21	24.8	39.2	51.7	47.9	48.5	51.7	63.0	56.2

(Table 4). The mean HCN content of the roots of each plant was nearly identical, i.e., 37.6 ± 13.7 and 35.4 ± 15.5 mg HCN/kg fresh weight. The coefficients of variation were 36 and 43%. The cyanide content of different roots from the same plant may vary by a factor of 4.

Table 4. Cyanide content in all roots of two different plants.

Root number	HCN (mg/kg fresh weight) Plant 1	Plant 2
1	29.3	29.1
2	30.9	36.7
3	35.1	26.0
4	30.2	34.2
5	54.2	43.5
6	53.7	30.6
7	55.7	44.5
8	50.9	82.6
9	51.3	41.4
10	16.6	26.5
11	31.6	29.7
12	44.1	42.8
13	18.9	18.1
14	24.4	23.9
15	-	21.4

At monthly intervals for 6 months, a root was removed from each of two plants and the cyanide content determined (Table 5). For plant 1, the mean was 37.6 ± 20.9 mg HCN/kg fresh weight with a coefficient of variation of 55.7%. For plant 2, the average was 85.7 ± 35.8 with a coefficient of variation of 31.8%. These results indicated an important variability, probably due to the fluctuations observed in roots from the same plant and differences linked to environmental conditions.

The HCN content was studied in 46 roots of various weights and lengths. The weight varied from 82 to 2083 g and the length from 10 to 62

Table 5. Variation in cyanide content of roots from the same plants during different months.

Month	HCN (mg/kg fresh weight) Plant 1	Plant 2
June	21.2	114.1
July	33.5	59.5
August	31.8	124.5
September	13.3	42.5
October	65.3	114.6
November	60.2	58.9

cm. The hydrocyanic acid content of the different roots fell between 9 and 162 mg HCN/kg fresh weight. There was no significant correlation between the weight or the length of the roots and their HCN content.

The analysis of the cyanide content of food items, at different stages of preparation, provided information on the effectiveness of detoxification (Table 6). The results indicated that the processes eliminate the majority of the glucoside; however, one cannot exclude, with the exception of foods cooked in oil, that the results cited are solely due to the destruction of the enzyme without complete destruction of linamarin, which is stable up to 150 °C (Cerighelli 1955). In the absence of the enzyme, linamarin becomes nondetectable during autolysis but could be catabolized by other pathways. Divergent views have been expressed on this subject (Rajaguru 1975; Jansz et al. 1974b). According to Pieris and Jansz (Jansz et al. 1974b; Pieris and Jansz 1975), cassava flour prepared simply by prolonged drying in the sun, according to the procedure followed by numerous populations, may still contain large quantities of cyanogenic glucosides.

Table 6. Cyanide content at different stages of food preparation.

Stage of preparation	HCN (mg/kg fresh weight) Roots	Leaves
Fresh	79.8	167.4
Peeled and dried	158.5[a]	NA[b]
Ground	NA[b]	93.3
Flour	1.8	NA[b]
Cooked in oil	NA[b]	1.0

[a]Increase in HCN is probably due to dehydration.
[b]NA not applicable to method of preparation.

Action of linamarase

Many plants contain beta-glucosidases, but the capacity to hydrolyze linamarin is limited to certain varieties, particularly those containing linamarin and lotaustralin (Butler et al. 1965). Moreover, the work of Butler et al. (1965) has shown that beta-glucosidase prepared from seeds of flax, after purification, contains at least two types of enzymes — one hydrolyzed linamarin and the other compounds of the amygdalin type.

A method for obtaining linamarase from cassava was described by Wood (1966). The activity of the linamarase prepared in this way confirmed that acid hydrolysis alone does not account for the liberation of hydrocyanic acid from linamarin. In addition, attempts to concentrate this preparation did not significantly increase its activity. A systematic study of the distribution of enzymatic activity in the different parts of the cassava plants was performed by de Bruijn (1971), who found that the enzymatic activity was highest in juvenile tissues, leaves, and the peel of the roots. No correlation appeared to exist between the enzymatic activity and the glucoside content. The enzymatic activity responsible for the degradation of linamarin might have involved several enzymes.

We have prepared linamarase according to Wood (1966) and tested it on synthetic linamarin (Calbiochem). Fig. 5 shows the degree of hydrolysis of linamarin: 1.52 $\times 10^{-7}$M for two dilutions of the enzyme preparation, 1/10 and 1/100. These results indicated a low activity in the preparation resulting probably from destruction of the enzyme during transport of the cassava to Belgium.

Biochemistry of Cyanide and Thiocyanate

The ingestion of sublethal doses of cyanide activates the body's own mechanisms of detoxification that ensure the transformation of cyanide into less toxic substances, principally

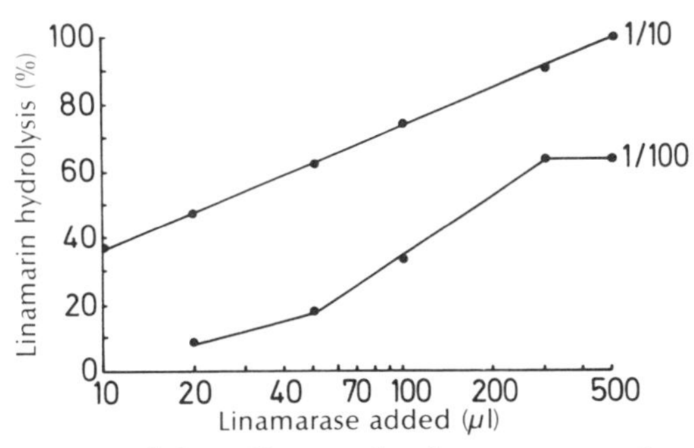

Fig. 5. Hydrolysis of linamarin by a linamarase extract from cassava peel.

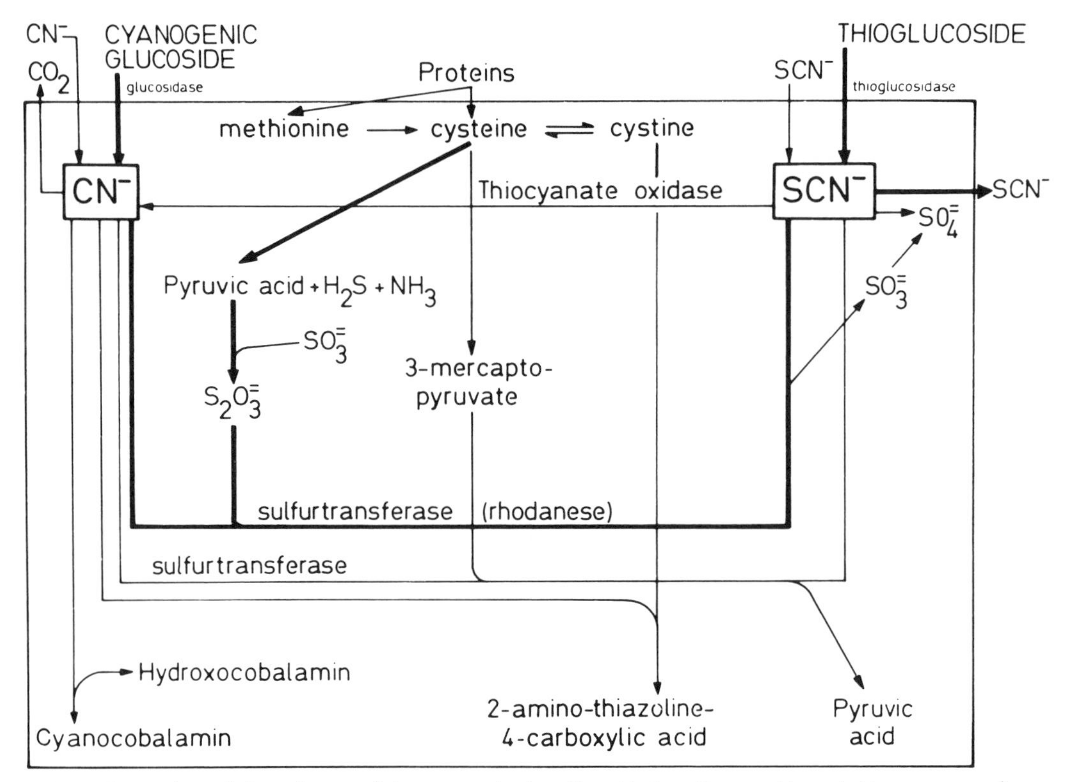

Fig. 6. Principal metabolic pathways of the cyanogenic glucosides, thioglucosides, cyanide, and thiocyanate according to Ermans et al. (1972).

thiocyanate. The various pathways for detoxification are shown in Fig. 6 and involve different enzymes and sometimes direct interaction with cellular components. The metabolic pathways are controlled by several compounds, including rhodanese, mercaptopyruvate-sulfurtransferase, cystine, and cyanocobalamin.

Rhodanese: The principal detoxification pathway is controlled by rhodanese (Fig. 7) (Lang 1933) or thiosulfate-cyanide-sulfurtransferase, which catalyzes the reaction by favouring the transfer of a sulfur atom, coming from a donor (thiosulfate, persulfide) to a nucleophilic accepter (cyanide ion). Thiosulfate ($S_2O_3^=$) is formed from sulfur-containing amino acids

$$S_2O_3^= + CN^- \xrightarrow{\text{rhodanese}} SCN^- + SO_3^=$$

Fig. 7. Principal metabolic pathway of cyanide with rhodanese.

whose presence is thus essential to the detoxification mechanism (Wheeler et al. 1975; Barrett et al. 1978). Rhodanese is widely distributed in the body, with the highest content being found in the liver and kidneys (Reinwein 1961; Auriga and Koj 1975). In the liver, the amount of enzyme varies from one species to another (Himwich and Saunders 1948) — a fact that may explain different susceptibilities to intoxication by hydrocyanic acid. Rhodanese intervenes actively to protect mitochondrial respiration from cyanide inhibition (Auriga and Koj 1975) as recently discussed by Westley (1973) in an extensive review of the problem. Thiocyanate, the product of this pathway of detoxification, is eliminated from the body mainly through excretion in the urine (Smith 1961).

Mercaptopyruvate-sulfurtransferase: 3-mercaptopyruvate, formed from deamination or transamination of cysteine, combines with cyanide by the action of a sulfurtransferase to form thiocyanate and pyruvic acid (Meister 1953; Fiedler and Wood 1956) as shown in Fig. 8. Like rhodanese, mercaptopyruvate-sulfurtransferase is present in numerous tissues, principally in the liver, kidneys, spleen, and pancreas (Meister et al. 1954).

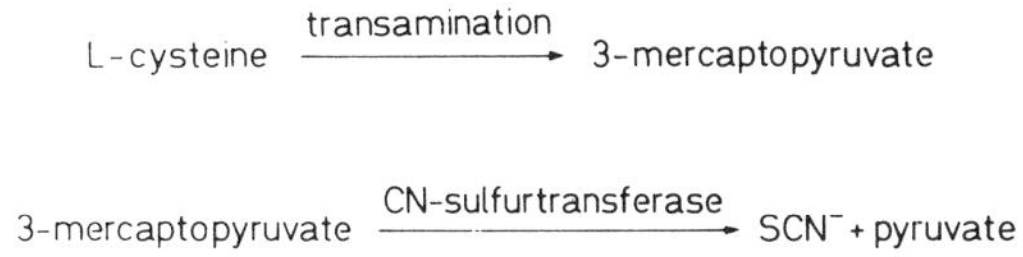

$$\text{L-cysteine} \xrightarrow{\text{transamination}} \text{3-mercaptopyruvate}$$

$$\text{3-mercaptopyruvate} \xrightarrow{\text{CN-sulfurtransferase}} SCN^- + \text{pyruvate}$$

Fig. 8. *Metabolic pathway of cyanide with 3-mercaptopyruvate.*

Cystine: Cystine is capable of reacting directly with cyanide, leading to a cleavage of the disulfide bond and formation of cysteine and beta-thiocyanoalanine (Fig. 9). Beta-thiocyanoalanine undergoes tautomeric conversion to 2-amino-thiazoline-4-carboxylic acid (Wood and Cooley 1956). By oxidative deamination, beta-thiocyanoalanine forms thiocyanopyruvic acid, which decomposes into thiocyanate (Schöberl et al. 1951); 2-amino-thiazoline-4-carboxylic acid can be excreted as such in the urine (Wood and Cooley 1956).

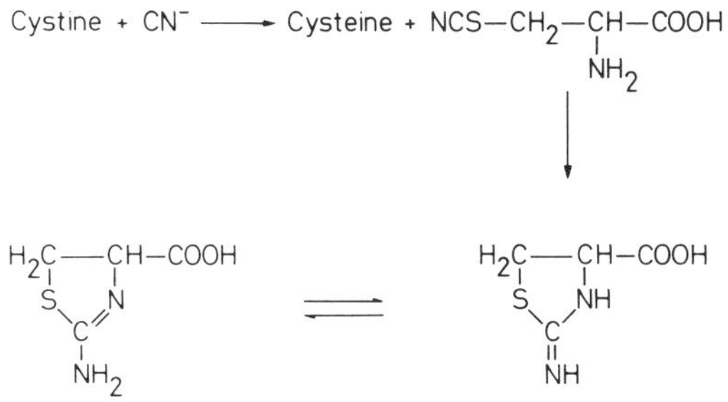

$$\text{Cystine} + CN^- \longrightarrow \text{Cysteine} + NCS{-}CH_2{-}CH(NH_2){-}COOH$$

H2C—CH-COOH (S, N, C=, NH2 ring) ⇌ H2C—CH-COOH (S, NH, C=NH ring)

Fig. 9. *Metabolic pathway of cyanide with direct action of cystine.*

Cyanocobalamin: Vitamin B_{12} or cyanocobalamin ensures an independent detoxification pathway by formation of hydroxocobalamin in the presence of light (Wokes et al. 1951). This hydroxocobalamin in turn reacts with cyanide to regenerate vitamin B_{12}. The importance of this pathway is illustrated by the increase in urinary thiocyanate in vitamin B_{12} deficiency (Smith 1961).

Other pathways: After injection of cyanide or thiocyanate labeled with ^{14}C, a fraction of the radioactivity is converted to $[^{14}C]CO_2$, the major part being excreted in the urine (Boxer and Rickards 1952). Approximately 45% of the injected isotope is recovered in the urine in the 24 hours following intravenous injection of $[^{14}C]$cyanide (Crawley and Goddard 1977).

In addition to the mechanisms for the detoxification of cyanide into thiocyanate it is important to note the presence in erythrocytes of a thiocyanate oxidase (Goldstein and Rieders 1953) that catalyzes the transformation of thiocyanate to cyanide. This process maintains the metabolic equilibrium between cyanide and thiocyanate in the blood. Under normal conditions, the ratio of plasma concentrations of cyanide and thiocyanate never excedes 1:99 (Pettigrew and Fell 1972).

Physiologic Effects of Cyanide and Cyanogenic Glucosides

Ingestion of cyanide or a cyanogenic glucoside can trigger diverse toxic manifestations.

HCN poisoning: The lethal dose of cyanide in adult humans is of the order of 50–60 mg (Halstrom and Moller 1945; Nicholls 1951) or 0.5 to 3.5 mg/kg body weight (Montgomery 1964). In animals, the lethal dose varies among species (Gibb et al. 1974); in cattle and sheep it is approximately 2 mg/kg body weight when absorbed in a single oral dose (Montgomery 1965). Several cases of cyanide poisoning have been reported in humans and animals (Montgomery 1964, 1965; Gibb et al. 1974) following ingestion of cyanogenic glucosides. Cyanide ion is rapidly absorbed from the gastrointestinal tract. It exerts its inhibitory action on many metalloenzymes. Specifically, cyanide forms a highly stable complex with cytochrome oxidase, thereby producing death by cellular anoxia.

Neurologic effects: Since the beginning of this century it has been known that repeated injections of cyanide can produce neurologic changes in animals (Swysen 1978). Repeated oral ingestion of sublethal doses can also lead to chronic neurologic problems (Wilson 1965).

In humans, two neurologic syndromes resulting from chronic exposure to cyanide have been recognized, Leber's disease and tobacco amblyopia (Monekosso and Wilson 1966; Anonymous 1969, 1970; Freeman 1969; Wilson et al. 1971), which are caused by hydrocyanic acid contained in tobacco smoke. These syndromes are probably due to a defect in the conversion of cyanide to thiocyanate (Wilson et al. 1971). They are characterized by an increase in serum cyanide and thiocyanate with a concomitant decrease in serum vitamin B_{12}. These observations confirm the metabolic interdependence of cyanide, thiocyanate, and vitamin B_{12} (Wilson and Matthews 1966; Wilson et al. 1971; Wells et al. 1972).

A syndrome called tropical ataxic neuropathy (TAN) has been observed in Nigeria (Osuntokun 1971; Osuntokun et al. 1968, 1969, 1970; Osuntokun and Monekosso 1969) and in Tanzania (Makene and Wilson 1972) where it is manifested almost exclusively in adolescents and adults. Although grouping of cases in families occurs, no hereditary factor has been demonstrated. The syndrome is associated with high cassava intake and protein malnutrition. There is a close correlation between the amount of cassava consumed and the increase in serum levels of thiocyanate and cyanide. The syndrome appears to be due to a defect in detoxification. Extremely low or unmeasurable serum levels of sulfur-containing amino acids (methionine and cysteine) have been observed. This probably results both from the intense utilization of these amino acids in the detoxification process and from the decreased intake of proteins and sulfur-containing amino acids. Discontinuation of cassava ingestion is followed by a decrease in thiocyanate levels, which rise again when the traditional diet is resumed.

Rats fed only on cassava for 18 months develop segmental demyelinating lesions of the sciatic nerve. Identical lesions have been observed in Nigerian patients with TAN. Moreover, 50% of the animals developed an ataxic neuropathy (Osuntokun 1970). These observations support the hypothesis that chronic cyanide intoxication of dietary origin (cassava) plays a role in the etiology of TAN in Nigerians.

Effects on thyroid function: Although the goitrogenic properties of plants containing linamarin have been known for more than 20 years (Care 1954), the goitrogenic effect of cassava was suspected for the first time in 1966 in East Nigeria where iodine deficiency alone could not account for the frequency of goitre (Ekpechi et al. 1966; Ekpechi 1967). The administration of cassava to rats produced an increase in thyroid weight, an increase in thyroid uptake, an increase in protein-bound iodine (PBI), a decrease in thyroid iodine reserves, and an elevated monoiodotyrosine/diiodotyrosine (MIT/DIT) ratio. This goitrogenic action of cassava was attributed to a thionamide compound (Ekpechi 1967). Shortly thereafter, Delange and Ermans (1971a; Delange et al. 1973; Ermans et al. 1973) ascribed the action of cassava to the antithyroid properties of thiocyanate formed from the catabolism of linamarin (chapter 2).

The metabolic abnormalities in the thyroid, induced in rats by cassava ingestion (Van der Velden et al. 1973), are identical to those produced by thiocyanate administration and are quantitatively related to the degree of endogenous thiocyanate production. Thus the antithyroid properties of cassava are linked to the endogenous catabolism of linamarin (chapter 8).

Therapeutic effects: Laetrile, structurally similar to amygdalin is used in Mexico and in 13 states in the United States for the treatment of cancer. Its use is highly controversial. A number of studies (Culliton 1973; Hill et al. 1976; Lewis 1977; Humbert 1977; Anonymous 1977a, 1977b; Wade 1977; Sadoff et al. 1978) in humans and in animals have given contradictory results. However, the majority indicate the total absence of therapeutic effect. Several instances of fatal poisoning have been reported after Laetrile injection.

Effects of Thiocyanate

Under normal conditions, the thiocyanate level in blood and urine results from protein catabolism. Thiocyanate concentrations are notably increased in smokers. Elevated levels may also result from the catabolism of thioglucosides present in certain plants, most of them belonging to the Cruciferous family (Michajlovskij 1964).

Chemically, thiocyanate behaves similarly to the halides (especially iodide) and, hence, has been considered "pseudohalide" (Wood and Williams 1949).

Thiocyanate is minimally catabolized under physiologic conditions and does not easily penetrate cells. It is concentrated in salivary and mammary glands and in stomach juice. The distribution space of thiocyanate approaches that of the extracellular space and, for a number of years was used routinely as a measurement of the latter.

The antithyroid properties of thiocyanate were first shown by Barker (1936) in patients with hypertension, treated with large doses of thiocyanate. These patients developed thyroid hyperplasia sometimes associated with signs of hypothyroidism. This observation was subsequently confirmed, and it was further demonstrated that the goitrogenic effect was prevented by the simultaneous administration of iodide or thyroid extract (Astwood 1943). The inhibitory action of thiocyanate on thyroid uptake has been clearly shown experimentally (Vanderlaan and Bissell 1946; Wolff et al. 1946; Wollman 1962) and is due to a specific effect on the mechanism of iodine concentration (Wolff et al. 1946). The existence of a thiocyanate concentration gradient in thyroid has never been demonstrated (Wolff et al. 1946; Vanderlaan and Vanderlaan 1947; Stanley and Astwood 1948; Wood and Williams 1949; Logothetopoulos and Myant 1956a; Wollman and Reed 1958).

In the thyroid, thiocyanate is rapidly converted to sulfate (Wood and Williams 1949; Logothetopoulos and Myant 1956a; Maloof and Soodak 1959). Thyroid-stimulating hormone (TSH) administration increases the intrathyroidal catabolism of thiocyanate (Sanchez-Martin and Mitchell 1960) and is even capable of reversing the block in uptake induced by this ion (Ohtaki and Rosenberg 1971). TSH probably accelerates the oxidation of thiocyanate to sulfate.

Halmi (1961), Wollman (1962), and Scranton et al. (1969) showed that thiocyanate, in low concentrations, inhibited iodide transport by increasing the velocity constant of iodide efflux from the gland. At high concentrations iodide efflux is greatly accelerated and thiocyanate inhibits the unidirectional clearance of iodide in the gland (Mitchell and O'Rourke 1960). At these high concentrations, thiocyanate also inhibits incorporation of iodide into thyroglobulin (Raben 1949). The competition between iodide and thiocyanate is at the level of the peroxidase (Maloof and Soodak 1959; Langer and Michajlovskij 1972).

The phenomenon of iodide–thiocyanate competition has also been studied in other tissues that concentrate iodide, i.e., salivary and mammary glands, stomach, and placenta (Halmi 1961). Thiocyanate and perchlorate inhibit nonthyroidal iodide transport but in different degrees in different organs and different species (Halmi 1961). Competition between nitrate, perchlorate, chloride, and thiocyanate ions for renal clearance has been observed (Walser and Rahill 1965).

The iodide and thiocyanate concentration in saliva and gastric fluid is superior to their plasma concentration (Logothetopoulos and Myant 1956a,b).

Prolonged elevation of serum levels of thiocyanate was studied by Funderburk and van Middlesworth (Funderburk 1966; Funderburk and van Middlesworth 1967, 1968, 1971) who showed that thiocyanate production diminishes in fasting rats but that, paradoxically,plasma concentration increases, probably because of a reduced rate of degradation and a decreased volume of distribution. In contrast, thiocyanate overload causes an accelerated disappearance of thiocyanate from the blood.

Action and Metabolism of Linamarin

The toxicity of cassava results from the degradation of linamarin and the concomitant production of cyanide. The effects of linamarin and the cyanogenic glucosides are in general still poorly documented. In sheep, lotaustralin has been shown to be toxic, and the lethal dose is estimated at 4–5 mg/kg body weight (Coop and Blakley 1949). Recent studies of linamarin (Barrett 1976; Barrett et al. 1977) have shown that:

• A dose of 50 mg is lethal in the rat. It is not clearly established whether the toxicity is only due to the production of hydrocyanic acid or whether linamarin is itself toxic.

• In the rat, 30 mg of linamarin produces a reduced thyroid radioactive iodine uptake and increased serum and urinary levels of thiocyanate, indicating an active detoxification mechanism. Linamarin is recovered unchanged in the urine and hence, can be absorbed as such from the digestive tract. However, no linamarin could be detected in the blood.

• The administration of linamarin to rats on a diet poor in sulfur-containing amino acids results in lower serum and urinary thiocyanate levels than those observed in animals receiving a normal diet and the same dose of linamarin. This is supplementary evidence for the participation of sulfured amino acids in the conversion of cyanide to thiocyanate.

The influence of linamarin on thyroid function has been investigated in our laboratory (Bourdoux, P., unpublished observations):

• Linamarin, 4–16 mg, was administered to rats on an iodine-poor diet, and its action was compared with that of thiocyanate administered under the same conditions (Table 7). The serum and urinary thiocyanate levels were not modified by linamarin administration. Thyroid iodide uptake was moderately reduced but notably less than after the administration of 2.7 mg of thiocyanate. The same experiment was repeated, but the animals were also given small amounts of cassava containing less than 0.3 mg cyanide but, presumably, containing the enzyme responsible for the hydrolysis of linamarin (Table 8).

Table 7. Effect of linamarin on iodine uptake and thiocyanate concentrations in rats 6 hours after administration.[a]

Rat	Drug (mg)	^{125}I uptake (% of dose)[b]	Serum SCN (mg/dl)[b]	Urinary SCN (mg/dl)[b]
1	–	21.3	0.70	0.22
2	NaSCN (1.7)	12.1	1.38	0.30
3	Linamarin (3.9)	11.7	0.73	0.20
4	Linamarin (7.8)	18.4	0.50	0.19
5	Linamarin (15.6)	16.4	0.88	0.19

[a]Each rat received a controlled diet of Remington ad libitum + KI (6.5 μg/100 g/d).
[b]Measurements were taken when the rats were killed.

Table 8. Effect of linamarin on iodine uptake and thiocyanate concentrations in rats 24 hours after administration.[a]

Rat	Drug (mg)	^{125}I uptake (% of dose)	Serum SCN (mg/dl)	Urinary SCN (mg/dl)
1	–	21.5	0.7	0.1
2	Linamarin (6.2)	16.0	1.7	1.5
3	Linamarin (15.4)	15.8	1.7	2.8
4	Linamarin (30.8)	11.3	2.1	3.3

[a]Each rat received a controlled diet: Remington ad libitum + KI (6.5 μg/100 g/d). Rats 2, 3, and 4 received a small piece of cassava peel as a source of linamarin.

These investigations demonstrated that, in the rat, the ingestion of linamarin alone does not cause an increase in endogenous thiocyanate production after a delay of 6 hours. In contrast, thiocyanate production is increased 20–50-fold when linamarin administration is associated with ingestion of small amounts of cassava, the presumed source of linamarase. In the second experiment, the percentage inhibition of thyroid uptake increased as a function of the dose of linamarin. This increased inhibition was accompanied by an increasing serum thiocyanate concentration.

• The action of linamarin was tested in humans. In two volunteers, after ingestion of 11 mg of linamarin, thiocyanate levels were measured during the 24 hours following drug administration (Table 9). One subject, a heavy cigarette smoker, had an elevated thiocyanate level at the start of the experiment. No modification of serum or urinary levels was observed. It was not possible to determine whether the absence of an effect of linamarin was due to failure to catabolize the drug or administration of an insufficient quantity.

• The hypothesis that endogenous catabolism of linamarin can be induced by microorganisms of the intestinal flora was also tested. The question was raised whether the preparation of cassava destroyed the enzyme, linamarase, leaving the substrate intact. Such a

Table 9. Effects of a single dose of linamarin (11 mg) on thiocyanate concentrations in serum and urine of man.

Time	Serum SCN^- (mg/dl)	Urinary SCN^- (mg/dl)[a]	t-test
Before	1.04	19.4 ± 3.8	
After	1.10	22.4 ± 5.0	NS
Before	0.28	11.1 ± 0.8	
After	0.26	11.1 ± 0.7	NS

[a]Mean ± SD.

possibility was suggested by Cerighelli (1955) and could account for the high serum and urinary thiocyanate concentrations observed in subjects ingesting large quantities of cassava flour that, after complete autolysis, contained extremely low quantities of cyanide. Known quantities of linamarin were incubated in acetate buffer (0.1 M; pH 5.5) at 37 °C for 24 hours in the presence of diverse microorganisms: *E. coli*, *Clostridium*, *Proteus*, and *Klebsiella*. No cyanide liberation was observed except in the presence of *Klebsiella*. *Klebsiella* produced complete hydrolysis of linamarin (Fig. 10). All strains tested were active and the extent of hydrolysis depended on the number of colonies of *Klebsiella* present in the medium during incubation.

Conclusions

Cassava, whose economic importance is constantly increasing, is an essential foodstuff of tropical populations. Its high carbohydrate content and the relative ease with which it can be grown are its two major advantages. However, because it contains the cyanogenic glucoside, linamarin, cassava must be detoxified before it is eaten. Otherwise, the cyanogenic glucoside, chemically similar to the amino acids, is capable of liberating large quantities of cyanide by hydrolysis and may lead to fetal poisoning and other side effects. Populations consuming cassava employ different detoxification procedures; however, these procedures, or at least some of them, are not effective, and the food consumed may still contain significant quantities of cyanide.

Quantification of the cyanide content of cassava has demonstrated variations from species to species, variations among roots from the same plant, seasonal variations, etc., which are essentially due to environmental factors.

During this study, we tested the direct effect of linamarin on thyroid function. In the rat, administration of linamarin by itself produced no effect. However, in the presence of its specific enzyme, it provoked an increase in serum and urinary thiocyanate levels and a decreased radioactive iodine uptake by the thyroid. In humans, no significant modification of thiocyanate levels was observed after ingestion of small amounts of linamarin. The hypothesis of a nonspecific degradation of linamarin by intestinal bacteria was also tested, and it was found in vitro that *Klebsiella* produces complete hydrolysis of linamarin.

Examination of the major pathway of cyanide catabolism has demonstrated the production of compounds less toxic than cyanide. Most often this catabolism requires important quantities of sulfur-containing amino acids and leads to the formation of thiocyanate, whose antithyroid properties are well established. The pathways are clearly linked and an insufficiency in one of the substrates necessary for detoxification may lead to the development of serious pathologic conditions. Thus, in populations where cassava is the basis of the diet it has been shown that this food plays a decisive role in the etiology of two syndromes, tropical ataxic neuropathy (TAN) and endemic goitre.

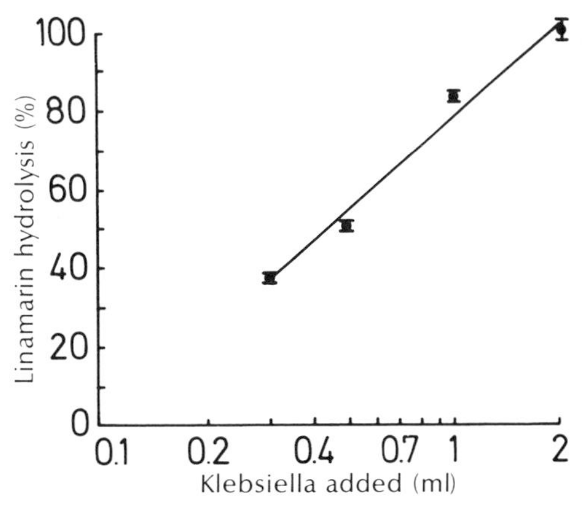

Fig. 10. Hydrolysis of linamarin by microorganisms (Klebsiella).

Chapter 2

Endemic Goitre in Kivu Area, Africa: Focus on Cassava

F. DELANGE, C.H. THILLY, AND A.M. ERMANS

Although it is generally accepted that iodine deficiency is the main etiologic factor in the development of goitre in isolated populations (Stanbury et al. 1954; reviewed by Delange and Ermans 1976), studies of endemic goitre on Idjwi Island (Delange 1966, 1974; Delange et al. 1968; Ermans et al. 1969; Thilly et al. 1972, 1973a), situated in central Africa near the eastern border of Zaire, have shown that iodine shortage by itself is insufficient to cause goitre endemia. All the inhabitants of Idjwi experience severe iodine deficiency, but only those living in well-defined regions suffer from high prevalences of goitre. The geographic distribution of goitre on the island indicates a relative failure of the populations in certain areas to adapt to some environmental factor closely related to the nature of the soil. Research suggests that cassava grown and eaten in the northern region harbours the goitrogenic factor (Delange and Ermans 1971a).

The population of this island, a mountainous mass roughly 50 km long and 10 km wide located in the middle of Lake Kivu, consists of about 40 000 Africans — all of the Havu tribe — living in 69 villages. Their dietary staples include bananas, beans, peanuts, sorgo, sweet potatoes, and cassava (Vis et al. 1969). They share the same primitive life-style and socioeconomic conditions and have little contact with the surrounding mainland.

The island is the seat of an extremely severe goitre endemia. A systematic epidemiological survey found the average prevalence of goitre on the island to be 32.3% (Thilly et al. 1972). However, goitre prevalence varied greatly by place, falling clearly into three distinct geographic areas: a hyperendemic region in the north, where the average prevalence of goitre was 52.8%; an endemic area of intermediate severity in the southeast, where average goitre prevalence was 26.0%; and a zone of very low endemicity in the southwest, where the average goitre prevalence was only 7.7% (Fig. 11). A preliminary survey, which involved 17000 inhabitants in 37 communities throughout Idjwi, had revealed goitre prevalence to be 10 times as great in the north as in the southwest (Delange et al. 1968).

Iodine Metabolism on Idjwi

The preliminary study also included comparative tests of iodine metabolism in the north and southwest regions of Idjwi. A random sample of male and female subjects 10–30 years old was selected for three measurements: the accumulation of radioactive iodine by the thyroid gland after the oral administration of 10 μCi of "carrier free" ^{131}I; the daily urinary excretion of stable iodine; and the level of PB^{127}I in the plasma. These parameters of iodine metabolism explored, respectively, the avidity of the thyroid gland for iodine, the level of alimentary iodine intake, and the level of circulating hormone.

In both goitrous and nongoitrous areas, thyroidal radioactive iodine uptake reached very high levels, whereas measurements of daily urinary excretions of stable iodine showed very low outputs (Table 10). [The levels of radioactive iodine uptake were slightly higher

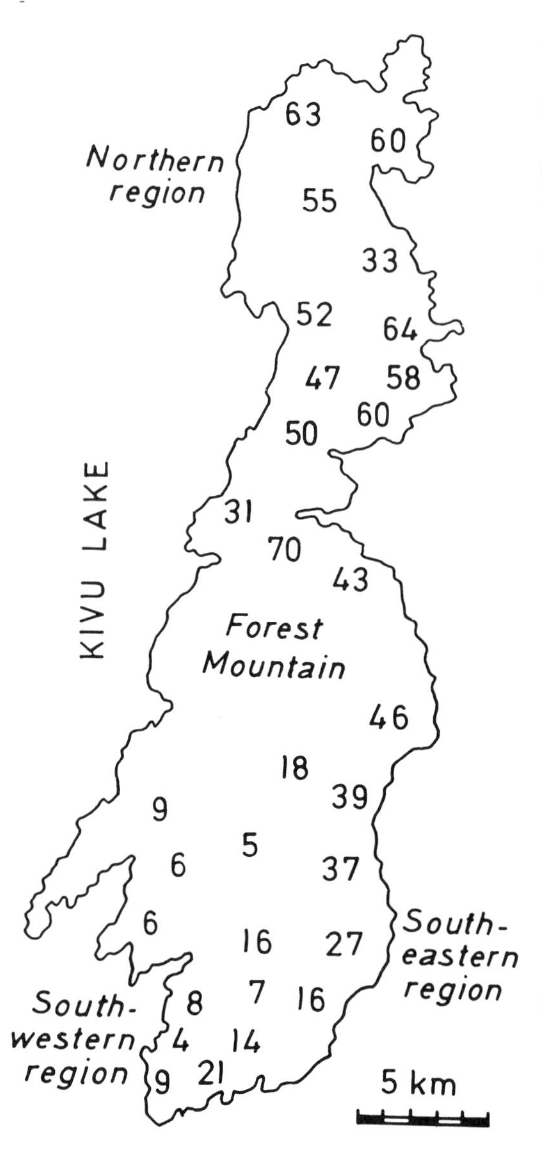

Fig. 11. *Goitre prevalence (%) in 30 villages of Idjwi island (Thilly et al. 1972. Amer. J. Clin. Nutr. 25, 30–40).*

in the north, but this might be related partly to the impairment of iodine kinetics in endemic goitre (Ermans et al. 1963).] A significant reduction of plasma PB^{127}I was observed in the goitrous area, compared with the figures in the nongoitrous area.

The concentration of iodine in water collected from rivers and other sources and used in various communities in the north and southwest sections of Idjwi was studied. Similar readings characteristic of inadequate iodine supply were found in both areas.

All the residents of two villages in the goitrous area received a single intramuscular injection of a slowly resorbed iodized oil. After the iodine therapy, goitre incidence regressed sharply (Delange et al. 1969; Thilly et al. 1973a; chapter 3).

Iodine Deficiency: a Permissive Factor in Endemic Goitre

In addition, the survey showed that, if the concentration of urinary iodine is read as an index of iodine intake, the whole population of Idjwi suffers from a severe and uniform iodine deficiency. In the southwest of the island, this deficiency is associated with only a very mild prevalence of goitre. In the north, however, it is related to severe goitre endemia, as shown by the high prevalence of goitre and cretinism (the survey revealed 89 cretins in the north, but none in the southwest) (Delange et al. 1972c) and by the decreased level of plasma PB^{127}I.

Although the curative influence of the iodized oil proved that iodine deficiency was necessary to allow the development of goitre, it did not explain the regional differences in goitre prevalence. Iodine deficiency in the north was moderately higher but a supplemental goitrogenic factor acting specifically in the northern region seemed likely.

The preliminary investigations were followed by a systematic epidemiological survey of Idjwi designed to look for ecological correlations between the variations in the prevalence of goitre, different environmental factors, and a series of parameters of thyroid function (Thilly et al. 1972).

A medical survey covered about 85% of the inhabitants in all 69 villages. Also included, as a control group, were the 1705 inhabitants of a community on the riverside of Lake Kivu in an area with no abnormal prevalence of goitre. Information was collected on each subject as to goitre type (Perez et al. 1960), age, sex, weight, height, name, parents, and code numbers of the family and village. The data permitted an

Table 10. Comparison of various parameters of iodine metabolism in a goitrous (region 1: north) and a nongoitrous (region 2: southwest) area of Idjwi Island.

Indicator	Goitrous area		Nongoitrous area		*t-test*
	Measurement	Number[a]	Measurement	Number[a]	(P value)
Goitre prevalence (%)	54.4	9000	5.3	3600	–
Cretinism prevalence (%)	1.0	9000	0.0	3600	–
Thyroidal ^{131}I uptake at 24 h (% of dose) (mean ± SE)	79.5 ± 0.7	376	70.8 ± 1.2	104	< 0.010
Daily urinary iodine output (μg) (mean ± SE)	13.1 ± 1.1	69	18.3 ± 4.0	48	> 0.100
Urinary ratio — iodine/creatinine (μg/g) (mean ± SE)	22.2 ± 1.3	56	29.6 ± 3.7	20	< 0.050
Plasma PB^{127}I (μg/100g) (mean ± SE)	3.4 ± 0.1	115	4.2 ± 0.1	40	< 0.001
Iodine content of water (μg/l) (mean ± SE)	4.2 ± 0.5	26	5.7 ± 1.2	10	> 0.100

Source: Delange, F. et al. 1968. J. Clin. Endocrinol. Metab., 28, 114–116.
[a]Number of cases or size of sample.

analysis of goitre prevalence as a function of place, age, and sex in the total population, as well as a study of the familial aggregation of goitre.

Goitre as a function of place: The survey of goitre prevalence confirmed the findings of the preliminary research. In the three regions, prevalence varied considerably from one village to another. The degree of endemicity ranged from 2% of the population of the control village to 70% of the residents of a community in the northern region.

Goitre as a function of age and sex: On Idjwi the prevalence of goitre is strongly dependent upon age and sex. In both men and women, it increases in the same way from birth to adolescence when it reaches a modal peak. In the northern area, goitre prevalence attains 80% in men and 90% in women aged 15–20 years. After age 20, it decreases rapidly in men, but much more slowly in women, in whom it remains at a high level during most of adulthood.

Familial aggregation of goitre: During the medical census, it was often observed that in the same villages there were whole families with goitres and others with none. This familial tendency to develop goitres was studied quantitatively in 10 villages in which goitre prevalence was relatively homogeneous. Measurements were made of the proportion of goitrous children in families in which both parents were goitrous, in families in which only one of the parents was goitrous, and in families in which neither of the parents was goitrous.

The results showed a definite association between parents and their children regarding the presence of goitre; however, the differences between children of goitrous and nongoitrous parents were slight. Among families of nongoitrous parents, the percentage of goitrous children was as high as 42.2% compared with 58.8% in the case of goitrous parents. Also, the presence or absence of goitre in a spouse was related to the thyroid size of the partner, although the association was weak.

The differences between the percentages of goitre observed in children of goitrous and nongoitrous parents were not clear-cut and were insufficient to explain the disease merely on the basis of monogenic heredity. This situation may correspond either to a form of multigenic heredity or to the intervention of etiologic factors of the microenvironment. The fact that the same degree of familial tendency was observed between children and parents as between father and mother, i.e., between subjects with no genetic ties, favours the second hypothesis.

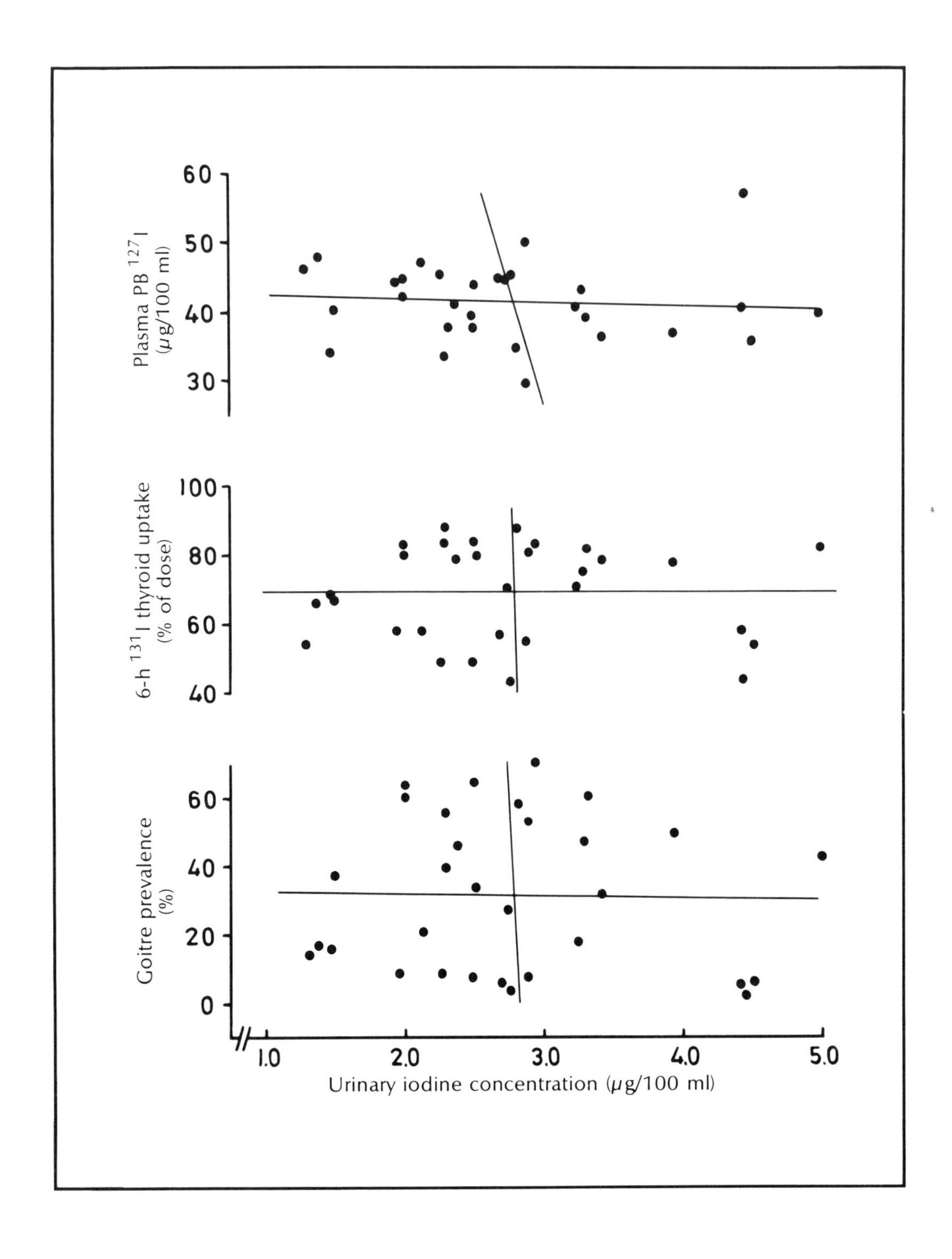

Fig. 12. Relationships between mean urinary iodine concentration and plasma PB^{127}I, thyroidal ^{131}I uptake at 6 hours, and the prevalence of goitre (Thilly et al. 1972. Amer. J. Clin. Nutr. 25, 30–40).

Goitre Prevalence and Iodine Metabolism

Relationships between regional goitre prevalence and parameters of thyroid function were studied in 693 subjects aged 15–19 years from 30 villages scattered throughout the island. The parameters of iodine metabolism investigated were the thyroidal uptake 6 and 24 hours after oral administration of 10 $\mu Ci^{131}I$, the level of plasma protein-bound iodine ($PB^{127}I$), and urinary iodine concentration.

The essential observation in the metabolic tests was the absence of systematic variations in urinary iodine concentration in relation to geographic factors, regional modifications in the prevalence of goitre, or ^{131}I thyroidal uptake. The relationships between mean urinary iodine concentration and variations of $PB^{127}I$, ^{131}I uptake at the 6th hour, and goitre prevalence are shown in Fig. 12. The data indicated clearly that the variations are unrelated to modifications of urinary iodine concentration. The concentration, which varied from 1.5 to 4.9 μg/100 ml, appeared to be stable and extremely low all over the island. That, as indicated in the preliminary tests, showed that iodine deficiency is homogeneous and severe throughout Idjwi.

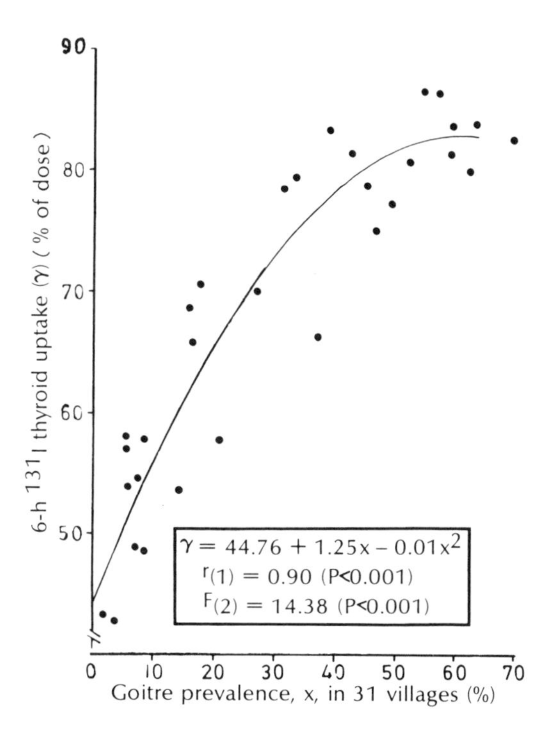

Fig. 13. Correlation between goitre prevalence and ^{131}I thyroidal uptake at 6 hours in the 31 villages of the study (Thilly et al. 1972. Amer. J. Clin. Nutr. 25, 30–40).

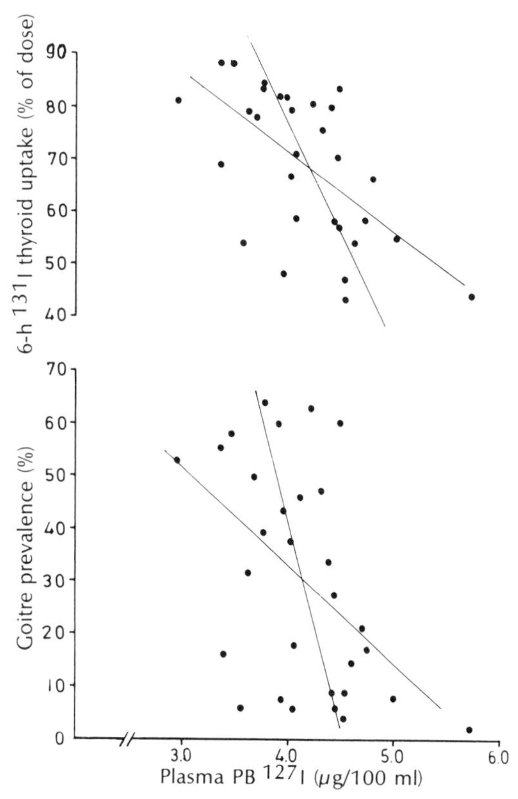

Fig. 14. Relationships between the level of plasma $PB^{127}I$ and the 6-h ^{131}I thyroidal uptake and the prevalence of goitre (Thilly et al. 1972. Amer. J. Clin. Nutr. 25, 30–40).

The prevalence of goitre, ^{131}I thyroidal uptake at the 6th hour, and $PB^{127}I$ were compared two by two. The results showed that uptake, which increased from 43% in villages with low goitre prevalence to 88% in those with high endemicity, was closely related to goitre prevalence (Fig. 13). The data also showed that $PB^{127}I$ levels decreased considerably from non-goitrous villages to high endemicity ones (Fig. 14).

To distinguish the variations of metabolism due to regional variations from those due to the presence or absence of goitre in a subject, we calculated separate averages for goitrous patients and for nongoitrous ones. In goitrous adults, thyroidal uptake was higher and PBI lower than in nongoitrous adults. However, the modifications of thyroid metabolism were influenced far less by the presence or absence of goitre in a particular subject than by regional variations in the prevalence of goitre. In the north, the alterations of iodine metabolism were much more marked in both the goitrous and nongoitrous subjects than in the inhabitants of the nongoitrous area. The most serious alterations were a drop in PBI in the goitrous subjects. In this region, therefore, goitre appears to indicate a relative failure of adaptation to some environmental goitrogenic factor. In the southwest, adaptation to very severe iodine deficiency is achieved by means of an increase in iodide trapping: this hyperavidity is due to increased thyrotropic hormone (TSH) stimulation (Delange et al. 1971). This adaptation takes place without abnormal prevalence of goitre.

The pattern of goitre prevalence and the extent to which the different parameters of thyroid metabolism vary on Idjwi Island are essentially determined by a geographic factor. Present data do not indicate that the degree of iodine deficiency or a familial genetic factor plays any role in this geographic distribution.

The geographic distribution of goitre on Idjwi clearly follows the pattern of the geologic nature of the soil (Ermans et al. 1969). The north of the island, which has a high goitre prevalence, is made up of a precambrian layer of granite, gneiss, and ruzizi, whereas the southwest, a nongoitrous region, corresponds exactly to areas of basalt-based soil. That isolated nongoitrous areas in an endemic region can correspond to areas of basalt-based soil had already been noted (Wilson 1954), and it had also been shown that the nature of the soil might play a part in the concentration of certain goitrogenic substances contained in plants (Sedlak et al. 1964). Moreover, the goitrogenic action of cassava had been suggested by studies carried out on rats (Ekpechi et al. 1966), and a food inquiry showed that, whereas the inhabitants of both the goitrous and nongoitrous areas of Idjwi have similar diets, cassava is consumed in larger quantities in the northern region (Vis et al. 1969).

Search for a Dietary Goitrogen on Idjwi

The preliminary studies had given rise to the suspicion that a dietary goitrogen was at work on Idjwi. Thus, the objective of a further study was to try to detect antithyroidal activity in the foods grown and eaten on the island (Delange and Ermans 1971a). The foods considered were bananas, peanuts, pumpkins, and cassava. Rice imported from Europe was used as a goitrogen-free control food.

The study included 131 inhabitants of the goitrous region and 67 from the nongoitrous area. The subjects, selected at random from lists compiled during the medical census of the island, were children and adolescents aged 7–14 years and clinically euthyroid.

Table 11. Influence of the ingestion of different foods on thyroid uptake of radioiodine in the goitrous and nongoitrous areas of Idjwi island.

	Goitrous area				Nongoitrous area			
Food ingested	Number of subjects	Average quantity ingested (g)	24 h ^{131}I thyroid uptake (% of dose) (mean ± SE)	*t-test* (P value)[a]	Number of subjects	Average quantity ingested (g)	24 h ^{131}I thyroid uptake (% of dose) (mean ± SE)	*t-test* (P value)[a]
Rice (control group)	22	315	86.8 ± 2.4		22	505	73.5 ± 2.1	
Cassava	27	490	71.9 ± 2.4	< 0.001	20	475	74.8 ± 2.1	> 0.5
Bananas	10	675	85.4 ± 3.7	> 0.5				
Peanuts	10	165	92.5 ± 1.7	> 0.1				
Pumpkins	10	520	79.5 ± 3.5	> 0.05				

Source: Delange, F. and Ermans, A.M. 1971. Am. J. Clin. Nutri., 24, 1354–1360.
[a]Significance compared with controls (rice).

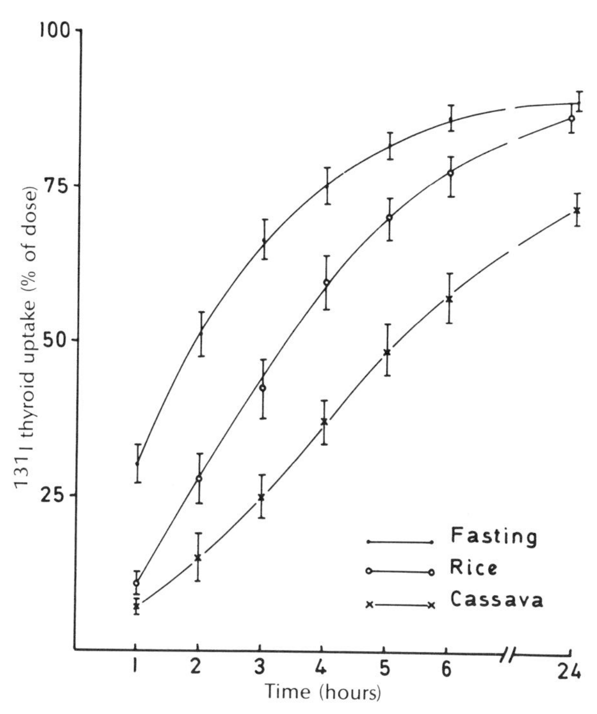

Fig. 15. *Pattern of the measurements of thyroid uptake of* ^{131}I *in fasting patients, the control group (rice meal), and the cassava group in the goitrous area of Idjwi island (Delange and Ermans. 1971. Amer. J. Clin. Nutr. 24, 1354–1360).*

Comparative studies were done to determine the influence of the ingestion of various foods on iodine metabolism, particularly on the distribution of a tracer dose of iodine-131. The research comprised three stages. The first consisted of a comparison of the influence of the ingestion of different foods on thyroid uptake of radioactive iodine among several groups of patients. Second, comparison was made of the urinary excretion of radioactive iodine and stable iodine after ingestion of the rice used as a control food and of cassava in two different groups of patients living in the goitrous zone. In the third stage, comparison was made of the curves of thyroid uptake of radioactive iodine obtained in the same patients after ingestion of a control meal and then after ingestion of a meal of local cassava.

The results of the tests are summarized in Tables 11 and 12 and in Fig. 15.

The study showed that ingestion of a meal of flour made from cassava grown in the goitrous area of Idjwi Island brought about a distinct drop in thyroid uptake of radioactive iodine. This effect was observed both in a comparison of several groups of patients who had eaten different foods and in a comparison of the same subjects taking successive meals of a control food and then cassava. In contrast, the absorption of cassava grown in the nongoitrous area of the island did not modify thyroid uptake to an appreciable extent, except in a few subjects.

The sequential measurements of iodine-131 uptake revealed a more or less marked delay in the accumulation of the tracer in the thyroid after the ingestion of the various meals. This delay is attributed to a slow-down in the digestive resorption of iodine-131 after the meal. If the mechanisms of thyroid and renal clearance are not affected by the nature of the meal, it must be expected that the final distribution of iodine-131 would be similar to what is observed when the tracer is administered to fasting subjects. This, in fact, is what was observed for all types of meals tested in the nongoitrous area as well as the goitrous area, with the exception of cassava grown in the north.

Furthermore, the reduction in thyroid uptake of iodine-131 induced by cassava from the north was associated with a distinct in-

Table 12. Comparison of urinary excretion of radioactive iodine and stable iodine in 11 subjects ingesting rice (control group) and 11 ingesting cassava in the goitrous area of Idjwi Island.

Iodine excreted in urine (0–6 h after food ingestion)	Control group (11)	Cassava group (11)	*t-test* (P value)
^{127}I (μg) (mean ± SE)	7.4 ± 2.2[a]	16.7 ± 1.8	< 0.005
^{131}I (% of dose) (mean ± SE)	5.83 ± 1.06	9.35 ± 1.67	> 0.05

Source: Delange, F. and Ermans, A.M. 1971. Am. J. Clin. Nutr., 24, 1354–1360.

crease in renal excretion of iodine-131. A significant increase in urinary elimination of stable iodine was observed under the same conditions. In a study that consisted of complete collection of urine samples with certainty during two periods of 24 hours each, it was found that after the cassava meal, the excretion of iodine-131 increased quantitatively in proportion to the reduction of the fraction of the dose fixed by the thyroid.

Goitrogenic Activity of Cassava

The studies showed, therefore, that a factor present in the cassava grown in the goitrous area of Idjwi is capable of inhibiting significantly uptake of iodine and increasing its excretion in the urine. This process is similar to that observed in humans who have eaten rutabaga (Greer 1962) or cabbage (Langer and Kutka 1964) or drunk milk from cows fed with goitrogenic foodstuffs (Clements and Wishart 1956). It is likely to play a causal role in the development of endemic goitre on Idjwi, in view of the low iodine intake of this population and the fact that cassava represents a large portion of the diet. The role played by cassava in the etiology of endemic goitre in Africa has already been suspected by other authors (Nwokolo et al. 1966; Ekpechi 1967; Oluwasanmi and Alli 1968).

It has been suggested that the goitrogenic activity of cassava is related to the presence of a thionamide-type substance (Ekpechi 1967). However, a more logical possibility is a thiocyanate-like substance. Cassava contains high concentrations of a cyanogenic glucoside, linamarin. This glucoside — when ingested — releases cyanide, which is catabolized into thiocyanate under the influence of rhodanese (chapter 1). The hypothesis that a thiocyanate-like substance is released from cassava is supported by several arguments: the similarity of the action of the cassava reported in the present study and that observed after large doses of thiocyanate (Wolff 1964); the high levels of thiocyanate in serum and urine of subjects in the goitrous area of Idjwi (Ermans et al. 1969); and the curative action of iodized oil (Thilly et al. 1973a), which is compatible with the presence of a goitrogen of the thiocyanate type but not with that of the thiocarbamide type (van Etten 1969).

In conclusion, the studies on Idjwi provide evidence that the absorption of cassava grown in the goitrous area of the island brings about partial inhibition of iodine uptake by the thyroid and an increase in its renal excretion. It suggests that this food, in the presence of an iodine deficiency, constitutes the goitrogenic factor. The mechanism of the action of the cassava and the reason that the same plant grown in other regions has no such action remain to be explained.

Chapter 3

Iodized Oil as Treatment and Prevention of Goitre in Kivu Area

C.H. THILLY, F. DELANGE, AND
A.M. ERMANS

Based on epidemiological and metabolic information (chapters 1 and 2), we concluded that iodine deficiency is only a permissive factor in the development of endemic goitre in north Idjwi. The underlying cause is goitrogenic substances whose action is possibly reflected in the serum thiocyanate levels that are five times those observed in Belgium. Our preliminary information suggested that these elevated levels came from ingestion of cassava. On the basis of our observations, two approaches to the prevention of this social plague were envisaged. One consisted in convincing the residents to modify their eating habits so that they decrease or eliminate the goitrogenic substance. In the long run, this approach is probably the solution to the problem of goitre in Ubangi but can only be proposed following extensive studies of dietary habits and the possibility of substituting other staple foods in the affected regions. A simpler approach consisted in studying whether an increase in iodine supply would prevent this endemic disease in spite of the continued presence of the goitrogens. The action of a series of goitrogens can indeed be mitigated by increasing iodine supply (Srinivasan et al. 1957; Greer 1962). This approach seemed particularly promising, as preliminary studies (McCullagh 1963; Buttfield and Hetzel 1967), carried out at the same time as we were studying endemic goitre in Idjwi, had shown the beneficial action of a new, slowly absorbable iodized oil. The injection of a single dose of iodized oil as a long-term source of iodine and protection against goitre seemed well adapted to the socioeconomic conditions of rural African regions and particularly to the isolated peoples on the island of Idjwi. The inhabitants live in an autosubsistence economic system in which the sale of commercial salt on the market is very limited and often replaced by potassium salts of plant origin (Pales et al. 1953; Vis et al. 1969).

The purpose of the work presented in this chapter was to evaluate the short- and long-term effects of iodized oil on a severe goitre and cretinism endemia whose epidemiological and metabolic characteristics are known. Specifically, we have studied:

- The effect of iodized oil on the prevalence of goitre in the general population after 6 months to a year (short term) and after 7½ years (long term);
- The duration of the effect of iodized oil on the prevalence of goitre as a function of age, sex, and type of goitre;
- The evolution and duration of iodine coverage by sequential measurements of urinary iodine excretion and thyroid uptake of ^{131}I;
- The reflections of this treatment on the principal parameters of thyroid function, i.e., iodine content of the gland and serum T_4 and serum TSH levels; and
- The incidence of cretinism and of hyperthyroidism by Iod-Basedow.

Experimental Protocol

The effects of iodized oil were studied in the northern part of the island of Idjwi. The evalu-

Table 13. Modification of thyroid size 1 year after the administration of iodized oil.

Type of goitre	Number of subjects	Thyroid size Unmodified (%)	Decreased (%)	Increased (%)
0	462	97.4	–	2.6
1a	95	9.5	87.3	3.2
1b	204	14.2	81.4	4.4
2	133	13.5	86.5	0.0
3	17	17.6	82.4	–

ation was divided into two sections:

- A longitudinal epidemiological study was carried out between January 1966 and August 1973 during which we measured the short- and long-term effects of one iodized oil treatment on the prevalence of goitre in the general population and as a function of age, sex, and type of goitre in two villages (Bugarula and Bukenge). In parallel, a metabolic investigation was performed in a limited number of treated young adults. In addition, the prevalence of goitre was measured at the beginning of the study in 1966 and 5 years later in a village in the north of Idjwi (Bukuruka) in which no treatment was administered. This village's population served as control.
- A cross-sectional metabolic study was undertaken in January 1971 during which we measured the principal parameters of thyroid function in limited groups of young adults in goitre-endemic villages of north Idjwi. The subjects differed in the lengths of time since treatment (1½, 3½, 5 years), and some had not been treated.

Epidemiological Results

Short-term effects: The results of iodized oil administration for the different categories of goitre, observed after 1 year, are presented in Table 13. Eighty percent of the subjects demonstrated a reduction in thyroid gland volume. This reduction was present both in the visible and voluminous goitres (stages 2 and 3) and in the smaller palpable goitres (stages 1a and 1b). No change occurred in approximately 10–15% of the subjects. An increase in goitre volume was noted in fewer than 5%.

The prevalence of goitre decreased from its initial value of 47% to 16% after 1 year. The reduction in overall prevalence was less marked in women than in men. This difference is probably due to the fact that the women had much larger goitres before treatment.

Table 14. Evolution of the total goitre prevalence after iodized oil treatment.[a]

Time after injection of iodized oil (y)	Male: Number of subjects	Male: Goitre prevalence (%)[a]	Female: Number of subjects	Female: Goitre prevalence (%)[a]	Both sexes: Number of subjects	Both sexes: Goitre prevalence (%)[a]
0	578	39.1	603	54.9	1181	47.2
1	426	9.4	462	22.9	888	16.4
3½	353	11.9 NS	408	20.8 NS	761	16.7 NS
5	324	21.9 ***	395	32.7 **	719	27.8 ***
6½	297	26.9 ***	352	36.1 **	649	31.9 ***
7½	224	25.0 ***	264	44.7 ***	488	35.7 ***

[a]Levels of significance of χ^2 calculated between the prevalence of goitre measured 1 year after the injection and at the time indicated: NS = not significant; * = $P < 0.05$; ** = $P < 0.01$; *** = $P < 0.001$.

Long-term effects: Table 14 illustrates the overall prevalence of goitre and the prevalence in each sex before and after treatment. In the population as a whole, the goitre prevalence decreased to 16%, 1 year after treatment, remained at this level 3½ years later, and then increased progressively after 5, 6½, and 7½ years, at which time the prevalence was still inferior to the initial value. The effect of iodized oil was similar in the two sexes, but goitre prevalence was greater at all times in women than in men. In the untreated inhabitants of Bukuruka, taken as control, the prevalence of goitre remained approximately the same during the 5 years of observation: 43% in 1966 and 47% in 1971.

Evolution as a function of age and sex: Fig. 16 shows the evolution of goitre prevalence in the same population as a function of age and sex. After 1 year, the prevalence had drastically decreased for all age groups and both sexes to a third or a fifth of its initial value. Later, there were notable differences in prevalence depending both on age and on sex.

In children, 0–4 years old at the time of treatment, the impressive drop observed after 1 year was followed by an increase that was clearly evident in both sexes 3½ years after treatment; 5, 6, and 7½ years after treatment, goitre prevalence in this age group was higher than the pretreatment value. At these observation times, the children were 5–12 years old, a period of life when goitre prevalence is particularly elevated in endemic regions. In children aged 5–14 years at the time of iodized oil administration, a slight resurgence of goitre appeared 5 years later. In this age group, 7½ years after treatment subjects were approximately 13–20 years old, when the prevalence pattern is different in the two sexes. In the young women, goitre prevalence continued to rise, whereas in young men it stabilized. In women between 15 and 39 years, the reappearance of goitre was slight and was only clearly manifest after 7½ years. In men older than 15

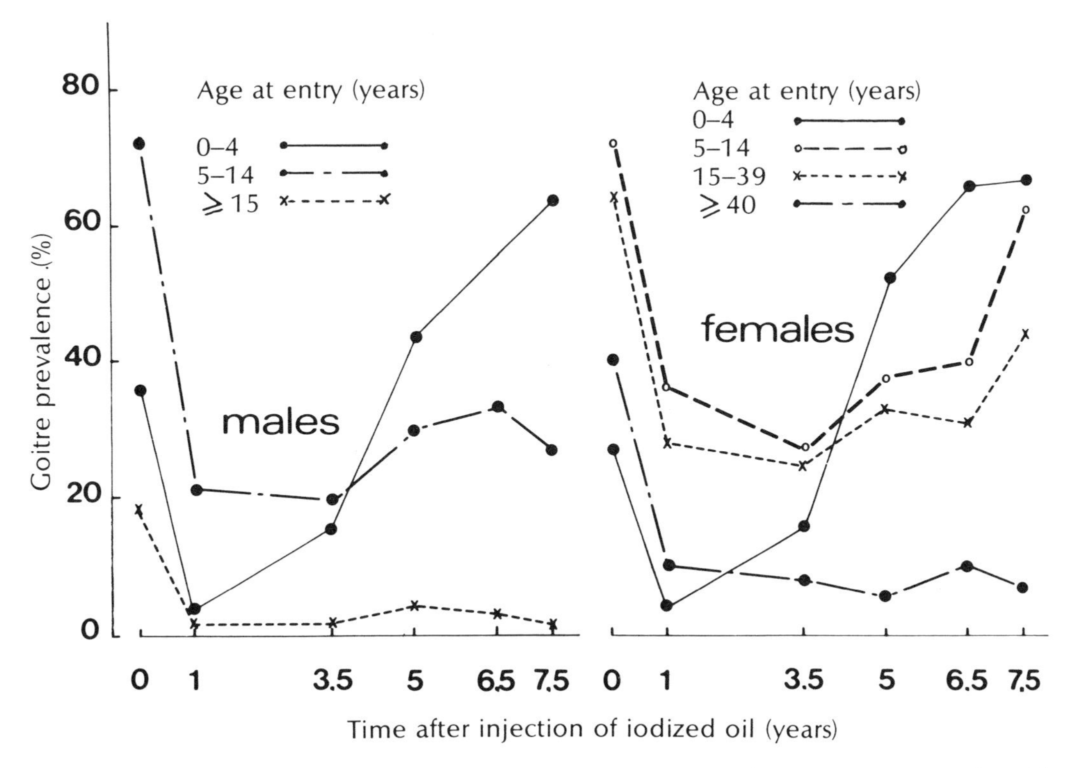

Fig. 16. *Evolution of goitre prevalence in subjects treated with iodized oil as a function of time after injection, age, and sex of the subject.*

Table 15. Evolution of various types of goitre after iodized oil treatment.[a]

Time after iodized oil treatments (years)	Male				Female			
	Number of subjects	Prevalence of goitre (%)			Number of subjects	Prevalence of goitre (%)		
		Palpable (stage 1)	Visible	Nodular		Palpable (stage 1)	Visible	Nodular
0	578	30.4	8.7	6.6	603	32.8	22.1	14.4
1	426	7.3	2.1	2.6	462	17.1	5.8	5.8
3½	353	10.2 NS	2.0 NS	2.0 NS	408	14.7 NS	6.1 NS	7.8 NS
5	324	19.8 ***	2.2 NS	1.9 NS	395	26.6 ***	6.1 NS	5.1 NS
6½	297	24.6 ***	2.4 NS	1.7 NS	352	30.1 ***	6.0 NS	7.4 NS
7½	224	22.8 ***	2.2 NS	1.8 NS	264	31.6 ***	12.5 **	12.1 **

[a]Levels of significance of χ^2 calculated between the prevalences of goitre measured 1 year after injection and at the time indicated; NS = not significant; * = $P < 0.05$; ** = $P < 0.01$; *** = $P < 0.001$.

years and women over age 40, the prevalence remained at the level attained 1 year after treatment.

Evolution as a function of the type of goitre: Table 15 summarizes the evolution in goitre prevalence as a function of the different categories of goitre, palpable goitres (stage 1), which are mild compared with the more severe types, visible (stages 2 and 3), and nodular. The drop in prevalence, observed in the short term in all groups, was also noted here. The table demonstrates particularly that in men there was no rebound in the prevalence of visible or nodular goitres during the observation period. In women the increase was only noted after 7½ years. The evolution of palpable goitres was similar to that described for all goitres.

Complications: No cases of hyperthyroidism, no local reaction, nor symptoms of toxic effect of iodine were observed. One abscess appeared but disappeared rapidly following appropriate antibiotic therapy. Surveillance of the population was sufficient to identify any cases of frank hyperthyroidism, although a small number of cryptic or temporary states of hyperthyroidism may have escaped observation. One case of cretinism occurred in the two villages receiving treatment during the first 5 years of observation, a child of a mother away from the village at the time of iodized oil administration. In the control village three new cases of myxedematous cretinism were detected during the 5 years following the first examination.

Metabolic Investigations

Correction of iodine deficiency: Table 16 shows the short-term evolution of urinary iodine excretion as a function of injected dose. Before treatment the urinary iodine concentration of approximately 2 μg/dl was the same in the three groups ($P > 0.05$). Six months later the concentration increased to 215 μg/dl in subjects having received 1000 mg iodine and to 134 μg/dl in those treated with 500 mg. After 1 year there was no statistically significant difference between the two treatment groups ($P > 0.05$). In the untreated subjects urinary iodine excretion was constant during the study period. In 24 urine samples containing important iodine concentrations, 96.5% of the iodine was retained by an ion exchange resin (Irosorb, Technicon) thus establishing the inorganic nature of the excreted iodide.

Fig. 17 presents the urinary iodine concentrations of the subjects receiving 1000 mg iodine and of the untreated subjects during the 7½ years of the study. The concentration fell from 215 μg/dl after 6 months to 82 μg/dl after 9 months, 55 μg/dl after a year, 13.8 μg/dl after 3½ years, and 7.5 μg/dl after 5 years. Even after 5 years, urinary iodine was still three times as high as in untreated subjects ($P < 0.01$). After 7½ years, the value in the treated subjects, 3.3 μg/dl, was no longer significantly different

Table 16. Modifications of urinary iodine concentration in respect to the time and the doses of iodized oil.[a]

	Before treatment		6 months after treatment		9 months after treatment		12 months after treatment	
Iodine dose (mg)	Number of subjects	^{127}I concentration (μg/dl)[a]	Number of subjects	^{127}I concentration (μg/dl)[a]	Number of subjects	^{127}I concentration (μg/dl)[a]	Number of subjects	^{127}I concentration (μg/dl)[a]
1000	55	2.0 ± 0.3 (0.6–4.6)	49	215.5 ± 23.6 (40.5–760.0)	45	82.3 ± 6.6 (19.0–185.0)	37	54.7 ± 5.9 (12.0–158.0)
500	19	1.9 ± 0.2 (0.3–3.0)	17	134.2 ± 44.7 (10.7–700.0)	17	63.0 ± 8.3 (13.0–135.0)	10	51.4 ± 17.8 (2.2–177.0)
0	19	1.7 ± 0.2 (0.4–4.1)	14	2.3 ± 0.2 (1.1–3.5)	14	2.0 ± 0.2 (1.2–3.0)	12	2.4 ± 0.2 (1.0–4.0)

[a]Mean ± SE; figures in parentheses are the range.

from the control value, 2.9 μg/dl. A semi-logarithmic plot of the decrease in urinary iodine excretion is not linear. However, one can trace two straight lines, one between 0 and 1 year where the half-life (t½) is 3.0 months,

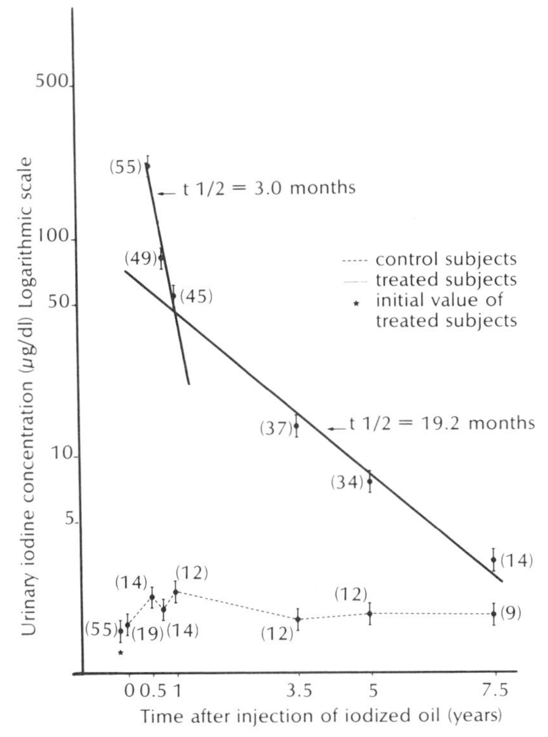

Fig. 17. Comparison of the evolution in time of urinary iodine concentration in the group of patients treated with 1000 mg of iodine with that of the group of patients treated with placebo. (Results presented as mean ± SE.)

and a second between 1 and 7½ years with a much more prolonged half-life of 19.2 months.

Evolution of thyroid function: Table 17 presents different parameters of thyroid function, measured in the cross-sectional study in the untreated group and in a group treated 1½ years earlier. In untreated subjects there was elevated thyroid uptake of ^{131}I, PB^{125}I at equilibrium, and serum TSH associated with low serum T_4 and an extremely low thyroid gland content of exchangeable iodine. In the treated subjects the normalization of serum thyroxine (10.3 ± 1.2 μg/dl) and of thyroid gland iodine content [7.2 (5.5–9.4) mg] accompanied a decrease in thyroid uptake of ^{131}I, PB^{125}I at equilibrium, and serum TSH.

Fig. 18 illustrates the evolution of goitre prevalence and the different parameters of thyroid function estimated just before treatment and at subsequent intervals (1½, 3½, and 5 years). Between 1½ and 5 years the serum T_4 and TSH values remained in the normal range, whereas iodine uptake increased progressively and thyroid iodine content fell; 5 years after treatment serum T_4 was 8.9 ± 0.8 μg/dl, serum TSH 3.6 (2.6–4.4) μU/ml, ^{131}I uptake 62.5 ± 4.7% at 24 hours, and thyroid gland content of exchangeable iodine 2.2 (1.6–2.9) mg.

Discussion

Evolution of goitre prevalence: The epidemiological analysis presented here shows that in severe endemic goitre, the administration of iodized

Table 17. Comparison of various parameters of thyroid function in untreated subjects and in subjects treated 1½ years before the determinations.

Parameter	Untreated subjects (n = 14)[a]	Subjects treated 1½ years before (n = 8)[a]	*t-test* (P value)
24 h ^{131}I thyroidal uptake (% of dose) (mean ± SE)	76.0 ± 4.0	37.5 ± 4.6	< 0.001
PB^{125}I at equilibrium	1.37 (1.15–1.63)	0.26 (0.18–0.38)	< 0.001
Serum TSH (μU/ml)	10.1 (7.6–13.3)	2.6 (1.9–3.4)	< 0.001
Thyroid exchangeable iodine pool (mg)	0.62 (0.53–0.73)	7.16 (5.45–9.40)	< 0.001
Serum thyroxine (μg/dl) (mean ± SE)	4.6 ± 0.5	10.3 ± 1.2	< 0.001

[a]Figures in parentheses are the range.

oil results in a drastic reduction in goitre prevalence. This decrease is present in all age groups, both sexes, and all categories of thyroid hypertrophy. The striking observation of the regression in volume of visible and voluminous goitres was unexpected but corresponded to results in other studies using iodized oil (Buttfield et al. 1967; Fierro-Benitez et al. 1969; Pretell et al. 1969a,b). This situation is in contrast to that observed after the introduction of iodized salt prophylaxis, where the regression has been slow and initially present only in the small goitres (Buzina 1970; Sooch et al. 1973; Mora et al.

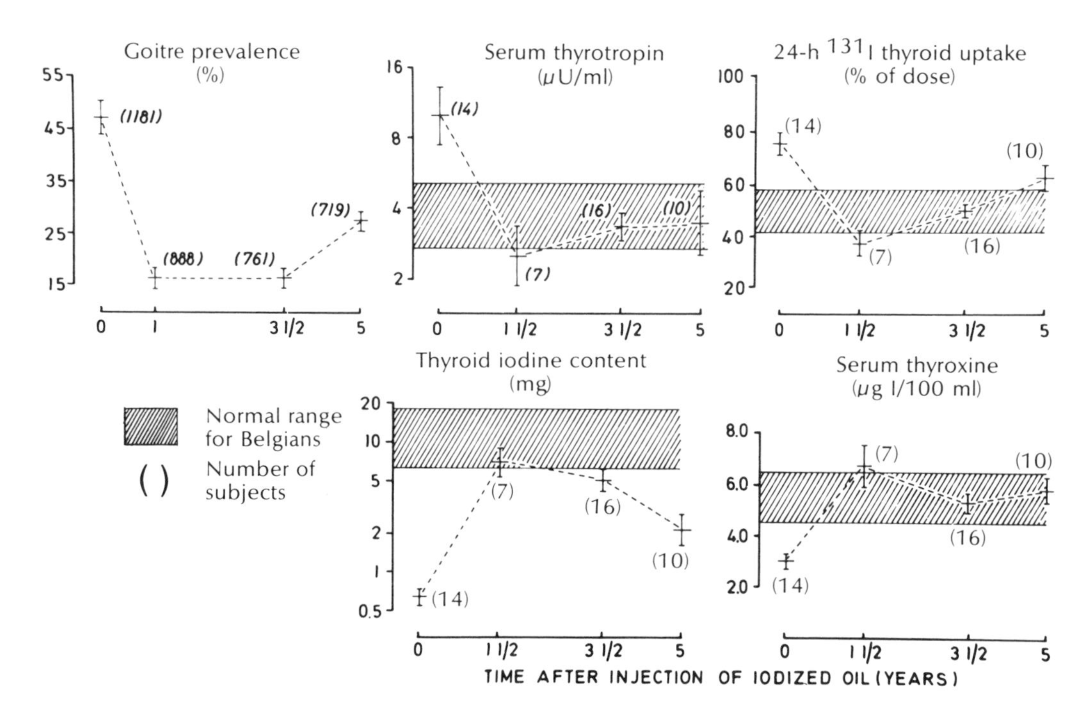

Fig. 18. Comparison of goitre prevalence and various parameters of thyroid function in untreated subjects and in groups of patients treated by iodized oil at various intervals of time before the determinations. (Results presented as mean ± SE.)

1974; De Leon et al. 1974). These results are in keeping with the observations of Marine and Lenhart (1909) who showed that iodine-deficient dogs, treated with physiologic doses of iodine, developed colloid goitres, whereas in those treated with very large doses of iodine, the hyperplastic glands reverted rapidly to normal volume.

The long-term study of the effects of iodized oil shows that goitre reappears much more slowly than initially suggested (Pretell et al. 1969a,b; Fierro-Benitez et al. 1969; Kevany et al. 1970; Stanbury et al. 1970) and is directly related to age and sex of the subject. In men over 15 years and women over 40, no reappearance of goitre in any category occurred during the 7½ years of observation. This suggests that the iodine load at these ages is sufficient to interrupt the self-perpetuating mechanism characteristic of goitrogenesis. In contrast, in the child, goitre reappears more quickly, 3½ years after treatment. This observation is similar to that of Pretell et al. (1969a) in Peru who showed a reappearance of thyroid hypertrophy 24 months after iodized oil treatment in the child compared with 42 months in the adult. It is also compatible with the report that of 15 cases of elevated serum TSH observed in the Himalayas 6 years after iodized oil treatment, 11 were in children (Ibbertson et al. 1974). These observations suggest the need for more frequent administration of larger doses of iodized oil in the child.

Correction of iodine deficiency: Before treatment, the urinary iodine concentration in young adults from north Idjwi Island was of the order of 2 μg/dl, which as discussed in the previous chapter, corresponds to a daily excretion of 16 μg. This indicates important iodine deficiency, found only in severe goitre endemia.

Between 6 months and 5 years after the administration of iodized oil, urinary excretion decreased from 2000 to 50 μg per day. The higher value is twice the daily excretion reported in certain communities in the United States (Oddie et al. 1970). The lower value is identical or slightly inferior to that obtained in European countries (Thilly et al. 1973b; Habermann et al. 1975). Analysis of the iodine excreted demonstrated that it was principally inorganic iodide and thus reflected accurately the quantity of iodine available for thyroid function.

The treated subjects' iodine excretion was rapid during the 1st year — a roughly biphasic curve with t½ = 3.0 months — followed by a much slower excretion (t½ = 19.2 months). This probably reflects the multiplicity of mechanisms involved in the handling of the administered iodine. Two at least may be cited: resorption at the site of injection (Buttfield and Hetzel 1969) and the probable formation of important iodized organic deposits in different tissues as shown by Costa et al. (1967, 1969) after administration of different iodized substances. An identical observation was reported by Pretell (1972) after administration of labeled iodized oil. It is possible that the rapid phase of elimination corresponds to the resorption of the iodized oil itself, whereas the slow phase results from the catabolism of the different iodized substances accumulated in the tissues.

During the first studies of iodized oil, the extrapolation of the urinary elimination rate based on the first part of the curve led to the prediction that the iodine load would be entirely eliminated approximately 18 months after treatment (Pretell et al. 1969a,b; Fierro-Benitez et al. 1969). The present observations, since confirmed by other studies (Pretell 1972), show a much more prolonged efficacy of this treatment. After 5 years, urinary iodine excretion was similar to that observed in Belgium.

The question must be asked whether the administration of such large doses of iodine perturbs thyroid activity by excessive accumulation of iodine in the gland or a blockage of uptake or organification. The normal serum T_4 levels and particularly the normal thyroid content of exchangeable iodine negate this hypothesis. On the contrary, these data, together with the results of other parameters 1½ years after treatment, demonstrate a complete normalization of thyroid function.

These findings are in agreement with the studies of Vagenakis et al. (1972), which showed that prolonged administration of elevated amounts of iodine to volunteers did not induce changes in circulating hormone levels, in contrast to the effect of smaller doses. The adaptive mechanism may correspond to a selective and direct block of active intrathyroid iodine uptake (Wolff 1976) without intervention of pituitary–thyroid feedback regulation.

Correction of peripheral metabolism and of iodine reserves in the thyroid gland: The serum levels of thyroxine and TSH and the other parameters of thyroid function, measured 1½ years after treatment, showed a complete normalization of iodine metabolism. Five years after the treatment they were still within entirely normal limits. From these observations we concluded that hormogenesis and the regulatory action of the pituitary were at their physiologic levels during this entire period.

It is remarkable that despite the very high quantities of iodine administered, the exchangeable iodine compartment of the thyroid gland did not exceed the limits of physiologic values but rather remained at the lower limit of normal. Probably for this reason this exchangeable pool was the first parameter of thyroid function to revert to abnormal values 5 years after treatment. This decrease and the concomitant increase in ^{131}I uptake were not associated with an elevated serum TSH. These observations are in agreement with the concept, noted above, that intrathyroid organic iodide, independent of TSH, can directly influence the degree of stimulation of the thyroid by an as yet unknown mechanism (Wolff 1976).

It is interesting to note that the reappearance of thyroid hypertrophy and the reduction in thyroid iodine reserves occurred at a time (5th year) when the urinary excretion of 50–60 μg was comparable to that observed in many nongoitrous regions, notably in Europe (Thilly et al. 1973b; Habermann et al. 1975). This observation supports the hypothesis that these glands are stimulated not only because of iodine deficiency but also by a distinct factor intervening at another level of iodine metabolism. As discussed earlier, this factor has been identified as thiocyanate coming from the cassava ingested in abundance by the population.

The appearance of a temporarily increased frequency of Iod-Basedow after the introduction of iodized salt programs is well documented (van Leeuwen 1954; Connolly et al. 1970; Stewart et al. 1971; Vidor et al. 1973). In contrast, no cases of Iod-Basedow have been observed in our study. This observation can be put in relation to the direct thyroid-blocking mechanism in the presence of important iodine excess cited above. It is known that this mechanism is not triggered by smaller quantities of iodine at the upper limit of normal such as in the programs using iodized salt. It is possible under these conditions to increase significantly the iodine content of the thyroid as well as thyroid hormone secretion. Another explanation would be that Iod-Basedow appears principally in subjects over age 50 who have multinodular goitres. The small fraction of the population in this age group and the spontaneous regression of goitre before this age could explain the absence of this complication in our study.

Chapter 4

Endemic Goitre and Cretinism in Ubangi

R. LAGASSE, K. LUVIVILA, Y. YUNGA, M. GÉRARD, A. HANSON, P. BOURDOUX, F. DELANGE, AND C.H. THILLY

Most of the studies presented in this monograph were carried out in the northwest of the Republic of Zaire, principally in the southern part of the subregion of Ubangi but also in the subregion of Mongala. These two subregions are situated in the region of the equator, which in the Republic of Zaire is located between 18° and 23° longitude east and between 1° latitude south and 5° latitude north (Fig. 19). This region covers approximately 300 000 km² and has a population of 2.5–3 million.

The subregion of Ubangi is bounded on the north and west by the Ubangi river, on the south by a line that is practically parallel to the equator and is an extension, from Mobeka, of the Zaire river. The east boundary is the Mongala river and a north–south line rejoining the Ubangi river. Geologically (Fig. 20), the north and central part of the subregion consists of precambrian terrain (Liki-Bembian and Ubangian); the southern part is alluvial phanerozoic terrain. The average altitude is 200–500 m. Two principal bioclimatic zones exist, separated roughly by the 4th parallel. In the north there is a plateau with subequatorial climate, the dry season lasting from December to February. The vegetation consists of savanna broken by forested bands (Fig. 21). The rest belongs to the central forest zone (Fig. 22), where the climate is equatorial with annual rainfall exceeding 1600 mm and falling throughout the year with two peaks at the equinoxes. The relative humidity is extremely high. The original vegetation consists of equatorial forest with undergrowth composed largely of ligneous plants rich in liana and epiphytes. Near dwellings, the forest has been extensively destroyed for agriculture (Fig. 23). Along the rivers and particularly in the angle formed by the Ubangi and Zaire rivers, the equatorial forest is flooded and the terrain swampy, only accessible by dugout canoe.

Administratively, the subregion of Ubangi is divided into five zones (Fig. 19): Bosobolo, Libenge, Gemena, Kungu, and Budjala.

Ethnography

The Ubangi populations are descended from Sudanese, Bantu, and Pygmy. There are six principal ethnic groups: Ngbaka, Ngbaka-Mabo, Mbanza, Mongwandi, Ngombe, and "Gens d'eau," the first four of whom speak Bantu. From an anthropological point of view, there are few or no differences among the six groups, which belong to the palenegrid system.

The geographic distribution of these groups is shown in Fig. 24. The Ngbaka are the main ethnic group and the only relatively homogeneous one. They occupy the zone of Gemena with extensions into the zones of Libenge, Bosobolo, Kungu, Budjala, and Karawa (subregion of Mongala). The Ngbaka-Mabo are a minor group located in the zone of Libenge, northeast of Zongo. The Mbanza are scattered in a number of zones around the Ngbaka: to the east of the Ngbaka of Mongala, in the north of Bosobolo, in the north and south of Libenge, and in the east and west of Budjala. The Mongwandi essentially inhabit the zone of Budjala with different enclaves, the most important of which is situated in the west of the zone of Kungu. The Ngombe are also scattered

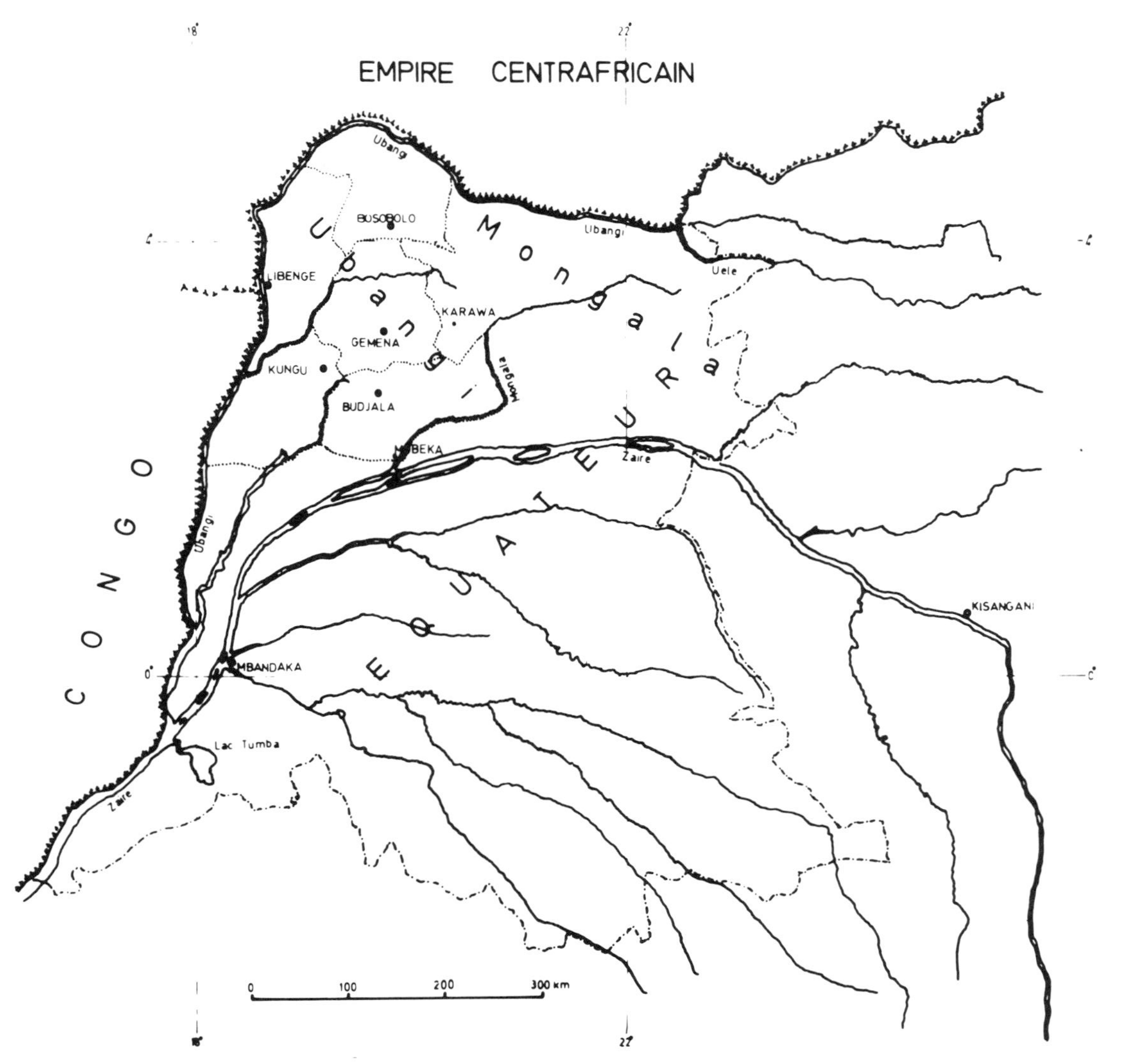

Fig. 19. *Region of the equator. The fine dotted lines indicate the limits of the subregion of Ubangi and of its five zones: Bosobolo, Libenge, Gemena, Kungu, and Budjala.*

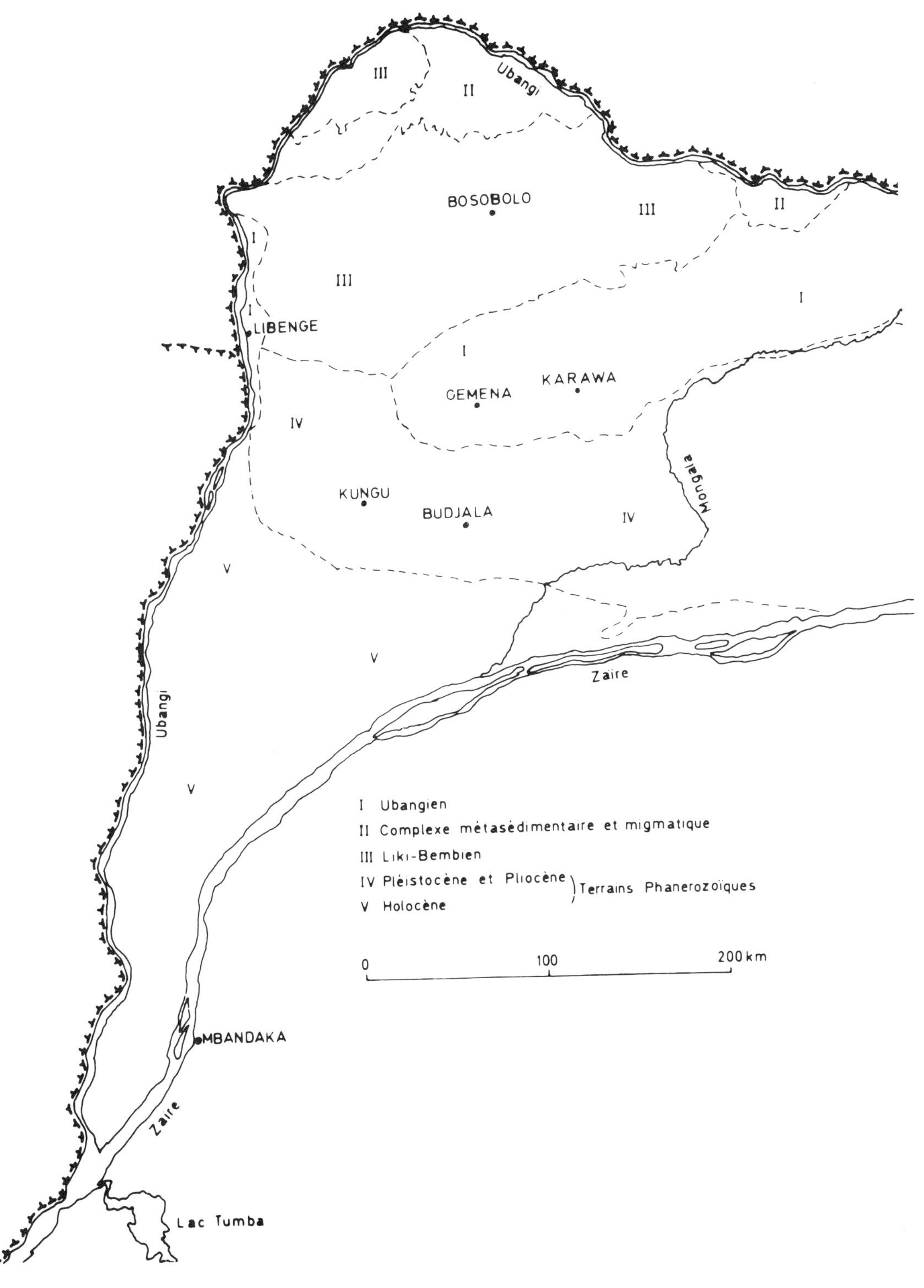

Fig. 20. Geology of the Ubangi area.

Fig. 21. Savanna broken by forested bands in north Ubangi.

Fig. 22. Equatorial forest in South Ubangi. One field is visible.

in several zones. They essentially occupy the region around Bosobolo as well as different enclaves in the zone of Budjala (northwest of the river Mongala) and Kungu (southeast). The "Gens d'eau" occupy the swampy regions situated between the Ubangi and Zaire rivers (Fig. 25).

These different ethnic groups are very similar culturally, particularly the Ngbaka and the Mbanza, although the different way of life of the "Gens d'eau," has produced a culture different from the land dwellers.

Family and Social Organization

The family and social organization is patrilineal with the exception of some Ngombe and "Gens d'eau," in whom traces of matriarchal organization persist. In the patrilineal system, the group comprises the descendants from a common ancestor on the male side. In such a system, the children belong to the father.

The husband lives in his own hut surrounded by those of his wife or wives. The huts are built of beaten earth, are round or rectangular, without windows, and with a thatched roof (Fig. 26). Coloured ornamentation is rare. The furnishings are very rudimentary consisting of a few chairs, mats, and receptacles for food.

There is no place for stocking food except for rare trestles.

The dwellings in a village are lined up along the paths (Fig. 26). The villages of the "Gens d'eau" have several thousand inhabitants, those of the Ngbaka 400–1000, and those of the Ngbaka-Mabo 300–400.

Economic System

The populations are essentially agricultural except the "Gens d'eau" whose principal activity is fishing. The economy is self-subsistent. Fields are laid out by burning the original forest, which disappears progressively around the villages. The forest fires, clearing of the residue, and the preparation of the soil is the men's work. The women are responsible for the rest of the work — planting, maintenance of the fields, harvesting, and gathering wood for domestic purposes. The fields are often planted with two or more crops. The tallest plants (corn, cassava) shield the lowest (peanuts). The soil is left fallow after 1–3 years, and new fields are prepared in the forest, often several kilometres from the village.

The crops vary from one region to another and particularly from one ethnic group to another. The Ngbaka, Ngbaka-Mabo, and the Mbanza cultivate mainly corn and cassava, the Mongwandi, the Ngombe, and the "Gens d'eau," cassava and plantain bananas. In addition, all groups grow peanuts, squash, yams, sugarcane, sweet potatoes, and fruit (papayas, mangoes, oranges). They ingest large amounts

Fig. 23. *Destruction of the forest near dwellings and paths to make way for crops.*

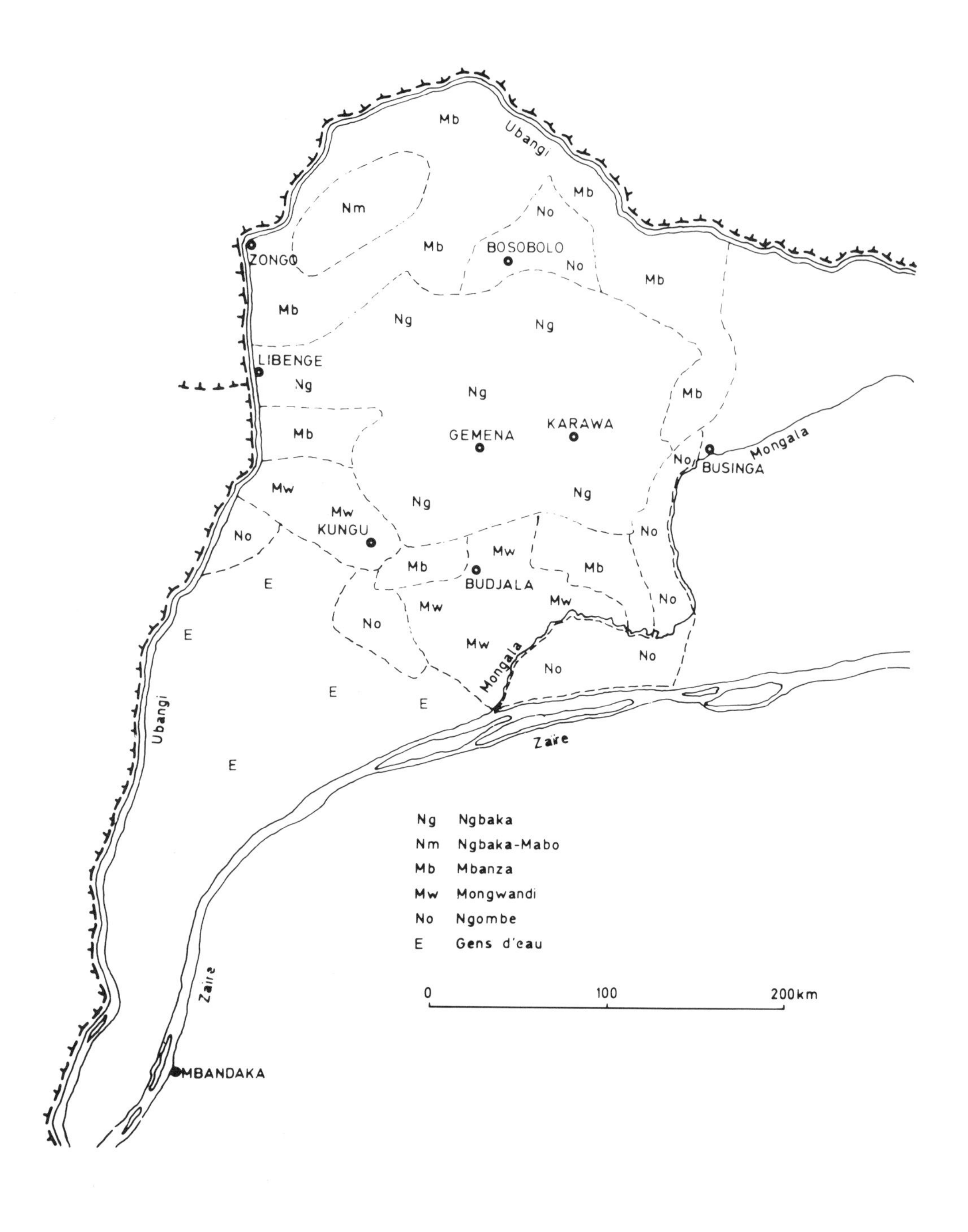

Fig. 24. Geographic distribution of the six principal ethnic groups in the Ubangi area.

Fig. 25. Village along river in the "Gens d'eau."

of palm oil. The Ngombe of the region of Bosobolo occasionally cultivate rice. Cotton and coffee are grown in addition to traditional crops and are destined for trade.

Hunting, once the main activity of the men, has diminished in importance because of the depletion of game, except in the savanna of the north Ubangi. The "Gens d'eau," Ngombe, and Mbanza fish with weirs, nets, and harpoons. The catch consists mainly of polypterus and silurus.

Only small livestock and poultry are reared (goats, pigs, chickens, ducks). The animals circulate freely in the villages and are a sign of wealth with greater exchange value than food value because they are only occasionally eaten. Other foods are gathered seasonally, insects (termites, ants), snails, caterpillars, and wild fruit between planting and harvesting.

More detailed descriptions of the area and its inhabitants are found in works by Guilmin (1922, 1933, 1947), Colle (1923), Tanghe

Fig. 26. Dwellings along the paths in the Ngbaka.

(1929a,b, 1930), Leontovitch (1933), Epoma (1949), Niçaise (1949), Maesen (1949), De Cocker (1950), Glorieux (1955), Burssens (1958), and Vansina (1966).

For a number of years the north and northeast of Zaire (Uélés) have been known to be regions of severe endemic goitre and cretinism (Kelly and Snedden 1962; De Visscher et al. 1961). Recently, the results of a preliminary study carried out in 1972–73 in the subregions of Ubangi and Mongala in northwest Zaire have been published (Thilly et al. 1977).

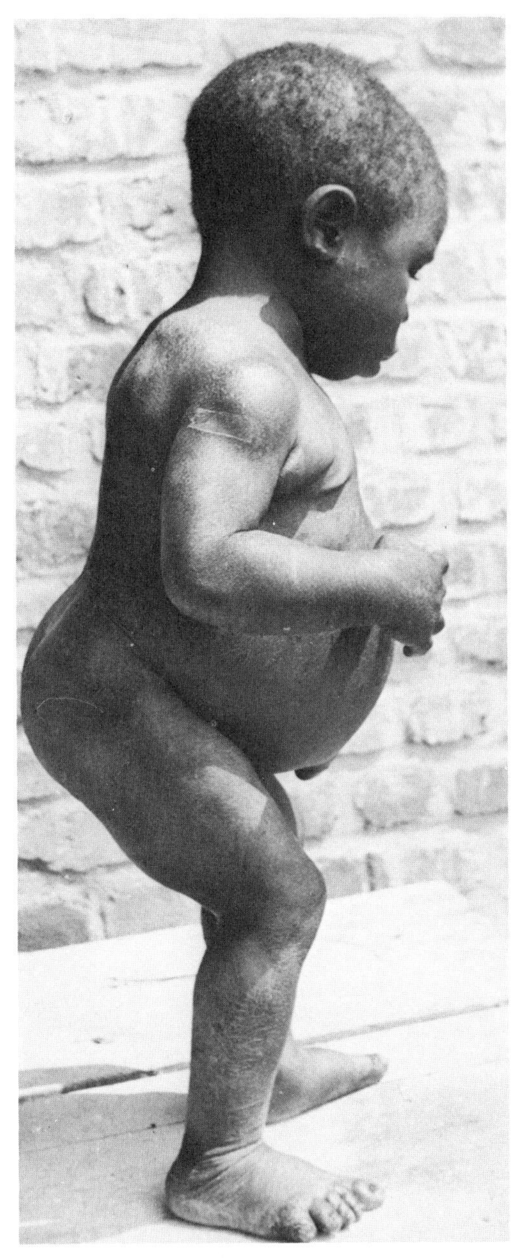

Fig. 27. Typical myxedematous cretin from the Ubangi area. This boy was 15 years old and his height was 82.5 cm.

This study describes in more detail the endemic goitre in Ubangi both epidemiologically and metabolically. In addition, the role of iodine supply and of thiocyanate in the etiology of endemic goitre in this region has been studied from a multifactorial point of view (Koutras 1974).

Materials and Methods

Ubangi has a surface area of about 65 000 km² with a population of approximately a million (Tshibangu 1978). The endemic goitre and cretinism observed in this region are among the worst in the world (Thilly et al. 1977). A preliminary survey in 1972–73 demonstrated the severity of this endemia, its geographic extent, and the frequency of alterations of thyroid function in the population. These preliminary results led to a mass treatment program with iodized oil (chapter 7) carried out by two mobile teams equipped with overland vehicles.

Patients: Our data were collected throughout the period of iodized oil treatment from October 1974 to December 1977. In all the villages visited, subjects receiving an injection of iodized oil were briefly examined: sex, age, volume, and nodularity of the thyroid were noted as well as the thyroid status of the subjects.

Endemic cretinism was assessed according to the criteria set out by the Pan American Health Organization (Querido et al. 1974): an endemic cretin is a subject born and living in an endemic area and exhibiting obvious mental retardation together with either a predominant neurologic syndrome (neurologic cretinism) or predominant hypothyroidism with stunted growth (myxedematous cretinism). Some subjects presented mild clinical signs of hypothyroidism or mental retardation or both; they have been considered as cretinoid (chapter 12 and Fig. 27–29).

Fig. 28. *Two myxedematous cretin girls, aged 9 and 7.5 years (on the right) and a cretinoid girl, 18.5 years old (on the left).*

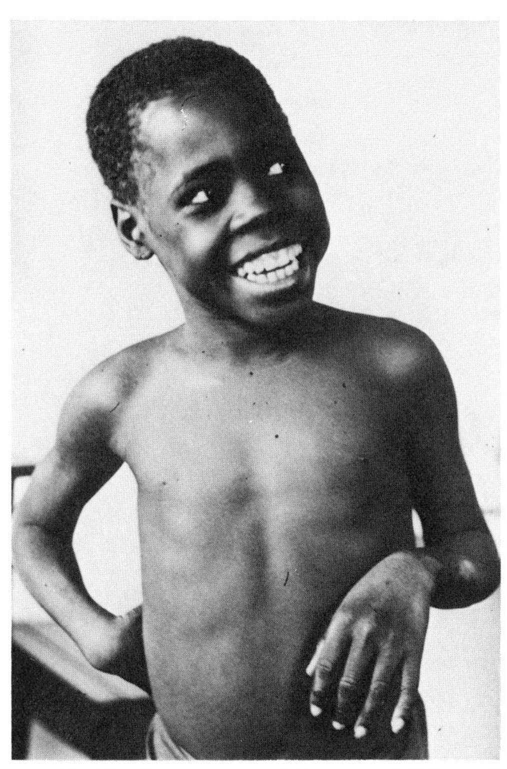

Fig. 29. *Neurologic cretin, aged 14 years and 115 cm tall. Spastic diplegia with hyperreflexia, strabismus, deafmutism, and severe mental debility.*

Information was gathered on precoded forms and thereafter on perforated cards for computer analysis. Our team examined 328 089 subjects, and the figures for prevalence of goitre and cretinism concern this entire population.

At the beginning of the mass treatment program, 5% of the villages had been chosen at random for a parallel evaluation of thyroid problems in the entire region. In each of these villages subjects of both sexes, aged 15–40 years, were examined, and their radioactive iodine uptake was measured. Single blood and urine samples were taken from each person for determination of serum concentrations of thyroid hormones, TSH, and thiocyanate (SCN) and urinary concentrations of iodide and thiocyanate. These subjects were all euthyroid clinically, with or without goitre. A map indicating the test villages for which complete biologic data are available is shown in Fig. 30.

Methods: We estimated thyroid volume, goitre stages according to the World Health Organization criteria established by Perez and colleagues (1962):

- Stage 0: no goitre; the thyroid is not palpable or the height of its lateral lobes does not exceed the length of the distal phalanx of the thumb of the subject;
- Stage 1a: palpable goitre; the goitre can be felt but is not visible with subject's neck in any position;
- Stage 1b: the goitre cannot be seen with neck in normal position but it becomes visible by extending the neck (Fig. 31);
- Stage 2: visible goitre; the goitre is visible with neck in normal position (Fig. 32);
- Stage 3: voluminous goitre; the goitre is large (Fig. 33) and is easily visible from a distance of more than 5 metres.

For the analysis of our results we considered the "all goitres prevalence rate" (AG) to be the percentage in the whole population of the goitres of any stage and "visible goitres rate" (VG) to be the prevalence of goitres of stage 2 and 3.

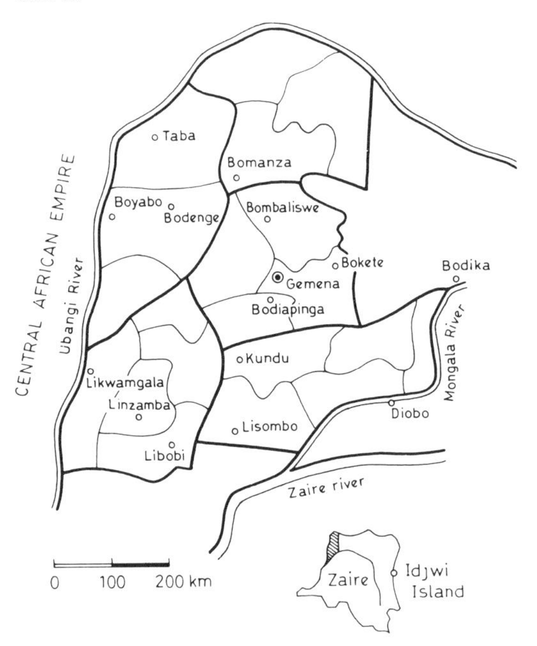

Fig. 30. *Map of Ubangi showing the administrative divisions and the locations of epidemiological and metabolic studies.*

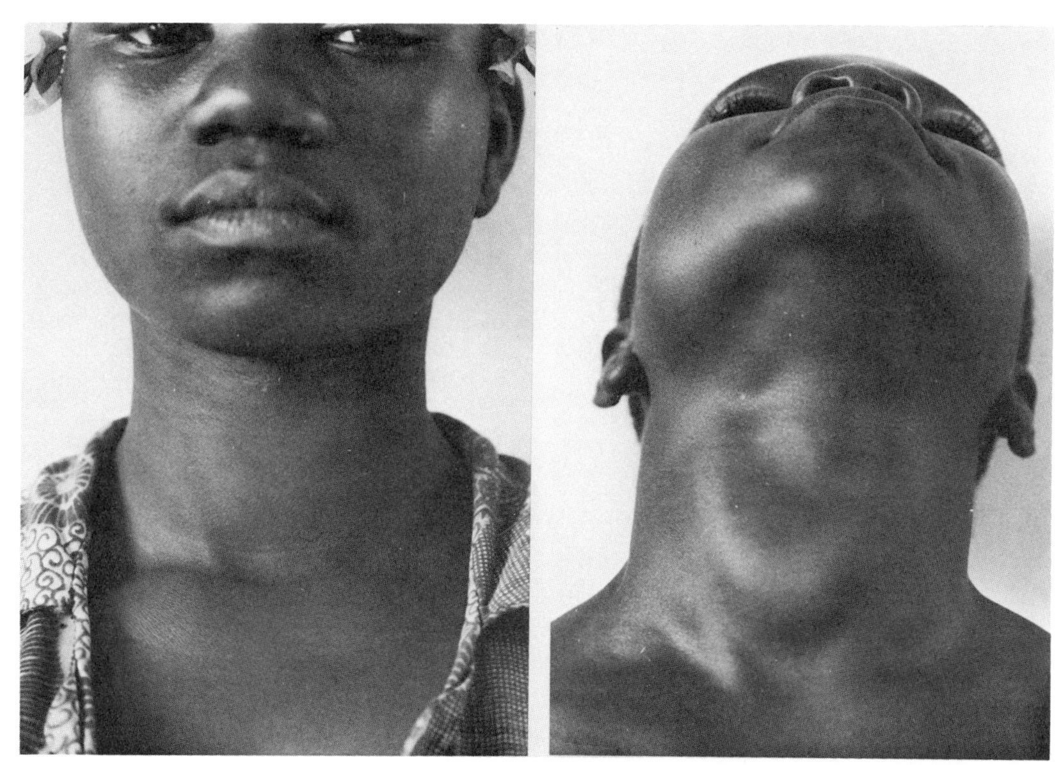

Fig. 31. *Goitre of stage 1b in a young woman. This goitre is well palpable and becomes visible by extending the neck.*

Different examiners were involved in this study; as new examiners entered the treatment program, they learned the procedures from their predecessors. Evaluations of thyroid volume were performed double-blind, and the maximal difference between observations of different investigators did not exceed 10% for goitres of any size (AG) and 5% for visible goitres (VG).

The methods for measuring thyroid function were 4-hour ^{131}I uptake by the thyroid; serum TSH, T_4, T_3, and thiocyanate concentrations; and urinary iodide and thiocyanate concentrations (Bourdoux et al. 1978).

The prevalence rates have been standardized for age and sex by a direct method of statistical analysis. The standard population used consisted of the 328 089 subjects examined in Ubangi between October 1974 and December 1977.

The values for serum TSH levels and urinary iodide concentrations have been considered as

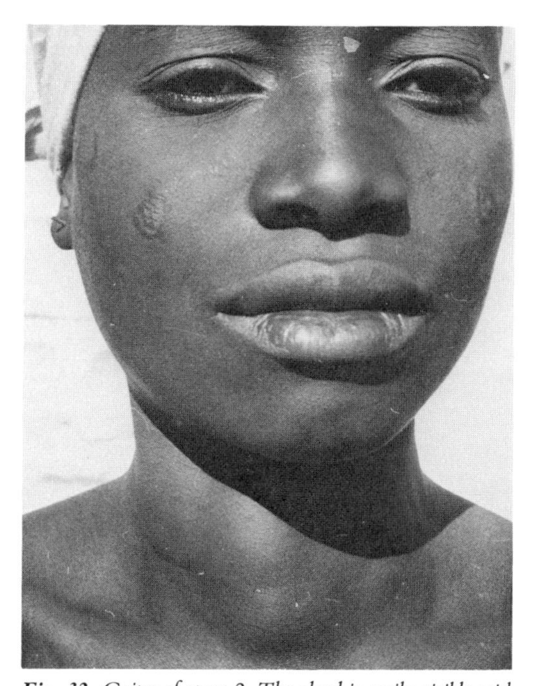

Fig. 32. *Goitre of stage 2. The gland is easily visible with subject's head in normal position.*

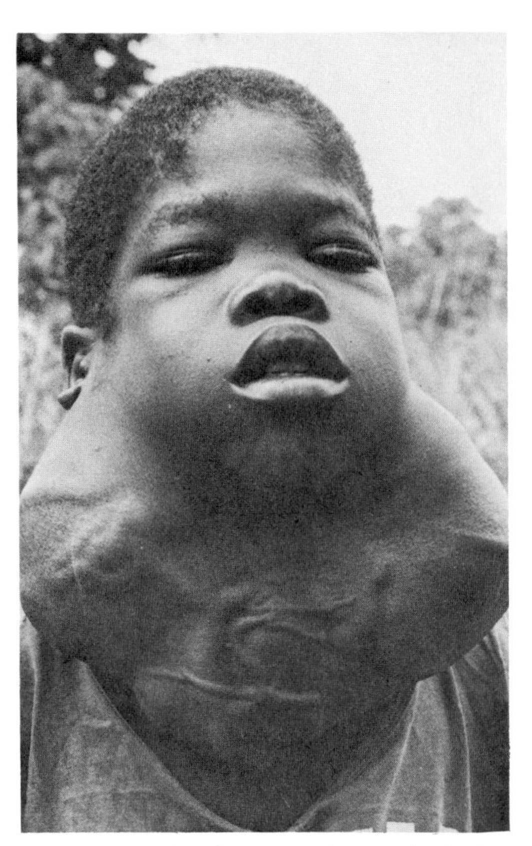

Fig. 33. Goitre of stage 3. Enormous hypertrophic gland in a boy, easily visible from more than 5 metres.

Results

For the population as a whole, the prevalence of goitre was 51%: 41% in males and 60% in females. Visible goitre was observed in 15%: 9% in males and 21% in females. The systematic examination of the population revealed 2358 myxedematous cretins (0.74%), 4137 suspected cretins (1.29%), 115 neurologic cretins (0.04%), and 86 deaf-mutes (0.03%). Cretinism was observed with equal frequency in males and females but rarely after age 25 because of the increased mortality associated with this condition.

We analyzed the data on prevalence of goitre according to age and sex (Fig. 34). For all goitres (AG) the maximal value in males (62%) was observed between ages 5 and 15. In females, the maximal prevalence was higher (76%) and occurred between ages 15 and 25. For visible goitres (VG) the maximal value in females (34%) was more than twice that in males (15%) and occurred in both sexes between ages 15 and 25. The prevalence of goitre decreased in both sexes after age 25. However the decrease was observed at an earlier age in boys, from age 15 onward.

Fig. 35 presents a map showing prevalence of goitre, visible goitre, and cretinism in the subdivisions of Ubangi corresponding to ap-

having a log-normal distribution. Geometric means have been calculated, and the coefficients of correlation calculated on the logarithms of the values. For the other results, arithmetic means ± 1 standard deviation (SD) are shown. The Student's *t-test* has been used for comparison of mean values where the F test of equality of the variances permitted it. Otherwise Cochran's *t-test* has been employed.

We applied a stepwise multivariate regression analysis to take into account joint effects of environmental factors. We analyzed the following variables without introducing any hierarchics: urinary iodide and urinary and serum thiocyanate in their simple form, and as the square root, the square, and the logarithm of the value. We retained in the final equation only the simple or transformed variables for the entering or removing of which the F test of Fisher attained the 5% level of significance.

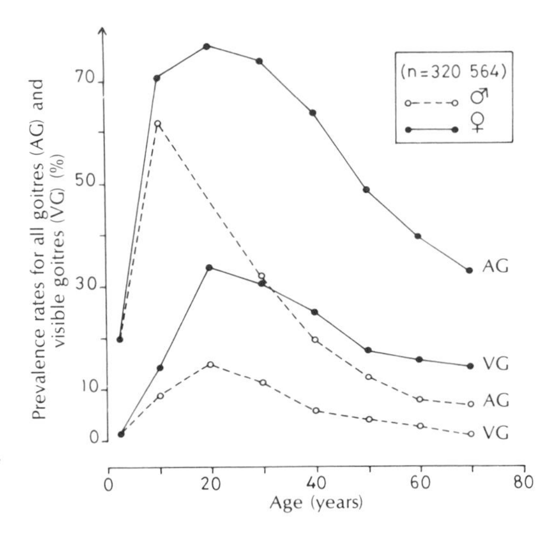

Fig. 34. Prevalence of goitre in Ubangi as a function of age and sex.

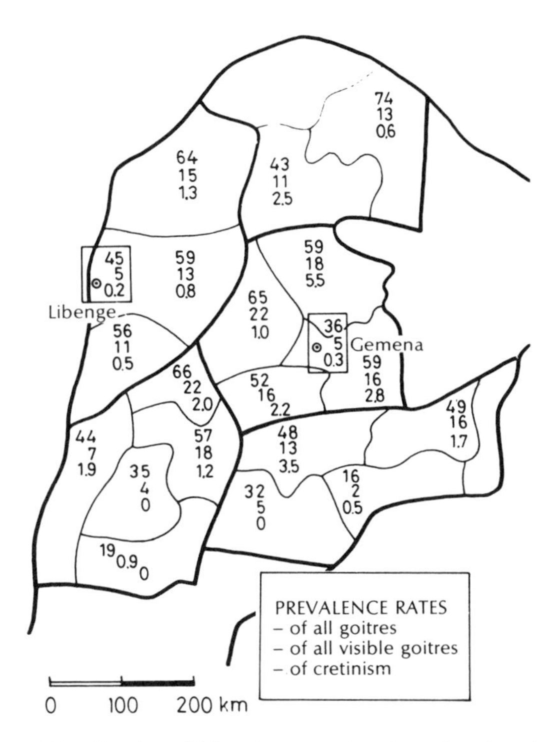

Fig. 35. Map of Ubangi summarizing the epidemiological data: prevalence of goitre of any size, of visible goitre, and of cretinism, expressed in percent, in each local subdivision. The values indicated in a square concern an urban population.

proximately 50 000 inhabitants. Because of the differences between the urban and rural environment (chapter 7), the values for Gemena and Libenge are indicated separately. The other results concern a 95% rural population. One observes the striking uniformity of values in the north and central parts of Ubangi and a progressive decrease in the five southern subdivisions. Cretinism is absent in the south, and the three measurements vary in parallel.

Table 18 presents the epidemiological data in 14 villages scattered throughout the territory. We have classified the villages in decreasing order of prevalence of goitre and grouped them into severely affected villages (VG > 10%) and mildly affected villages (VG < 10%). In severely affected villages, the prevalence of all goitres ranged from 48 to 74% (average 66%), of visible goitres 15 to 29% (average 25%). The prevalence of cretinism was particularly high, ranging from 1.7 to 9.0% of the total population. For the mildly affected villages, overall prevalence was 25%, visible goitres 2%, and cretinism 0.2%. The latter villages were located in the south and in the west near the Ubangi river. In three of the villages no subjects with visible goitre were observed and in five, no cretins. The biologic data collected

Table 18. Principal epidemiological data in endemic goitre in Ubangi collected in 14 rural villages.

Area	Number of subjects	Prevalence of goitre (palpable and visible) (%)	Prevalence of visible goitre (%)	Prevalence of cretinism (%)
Severely affected regions				
Taba	332	74	28	9.0
Bodenge	1048	72	27	3.2
Bodiapinga	1060	68	25	1.7
Kundu	900	68	29	6.0
Bombaliswe	453	66	23	3.8
Bokete	802	62	23	8.9
Bomanza	494	51	17	4.5
Bodika	222	48	15	2.3
Mildly affected regions				
Boyabo	1256	44	4	0.5
Diobo	113	24	9	0
Linzamba	773	23	2	0
Lisombo	220	23	0	0
Likwangala	445	18	0	0
Libobi	1013	7	0	0

Table 19. Principal laboratory data in endemic goitre in Ubangi in 412 young adults from eight rural villages in the severely affected zone.

Area	Number of subjects	4-h thyroid ^{131}I uptake (% of dose) (mean ± SD)	Serum concentration T_4 (μg/dl) (mean ± SD)	T_3 (ng/dl) (mean ± SD)	TSH[a] (μU/ml)	Urinary concentration SCN^- (μg/dl) (mean ± SD)	I^{-a} (mg/dl)	Urinary SCN^-/I^- (mean ± SD)
Taba	267	52 ± 13	3.7 ± 2.8	196 ± 56	8.7	3.08 ± 1.67	1.19	4.45 ± 5.03
Bodenge	42	-	5.6 ± 1.9	193 ± 35	6.0	1.30 ± 1.08	1.11	1.27 ± 1.32
Bodiapinga	20	55 ± 18	3.1 ± 1.7	136 ± 66	21.7	0.78 ± 0.43	1.26	0.57 ± 0.29
Kundu	36	-	3.7 ± 2.4	184 ± 51	7.1	1.70 ± 0.60	1.95	0.89 ± 0.39
Bombaliswe	11	49 ± 16	4.4 ± 1.5	174 ± 87	7.2	0.73 ± 0.40	1.47	0.57 ± 0.30
Bokete	33	65 ± 14	4.4 ± 2.8	142 ± 29	9.2	1.02 ± 0.68	1.03	0.87 ± 0.46
Bomanza	8	-	4.7 ± 1.5	162 ± 41	14.4	3.02 ± 1.55	2.40	1.20 ± 0.49
Bodika	13	52 ± 16	4.0 ± 2.1	110 ± 51	11.5	1.96 ± 2.15	2.41	0.86 ± 0.98

[a]Measurement is the geometric mean.

in these two groups of villages are presented in tables 19 and 20. In the severely affected villages 412 subjects were investigated; the prevalence of goitre among them was 68% and that of visible goitre 32%. The mean values were characteristic of particularly severe goitre endemia with very high 4-hour radioactive iodine uptake, serum T_4 concentrations at the lower limit of normal or below, TSH abnormally elevated, and T_3 levels normal or increased. These results were associated with very low urinary iodide and variously elevated urinary thiocyanate. The ratio of thiocyanate to iodide in the urine emphasizes the combined effects of these antagonistic elements.

In the mildly affected villages, 144 subjects were investigated, and in them the prevalence of goitre was 33% and of visible goitre 1.4%. The difference between the two groups of villages was highly significant ($\chi^2 = 73$; $P < 0.0001$). The mean values were comparable to those observed in Belgium for radioactive iodine uptake, T_4, T_3, and TSH. Compared with Belgian subjects, their urinary iodide concentration was slightly lower and thiocyanate level twice as high (Bourdoux et al. 1978). Compared with the severely affected villages, the urinary thiocyanate was twofold lower, and the urinary iodide was three times as high. The difference in mean values for each of the seven measurements of thyroid function in the two groups of villages was highly significant ($P < 0.001$). There was an overlap between the mean values of urinary thiocyanate and iodide in the two groups of villages, but the means of the thiocyanate/iodide ratio were clearly separated. In severely affected villages the mean ratios were all higher than 0.57, whereas in mildly affected villages they were lower than 0.47.

Table 21 presents the simple coefficients of correlation between the epidemiological characteristic (all goitres, visible goitres, and cretinism) and the biologic measurements

Table 20. Laboratory data measured in 144 young adults from six villages in the mildly affected zone.

Area	Number of subjects	4-h thyroid ^{131}I uptake (% of dose) (mean ± SD)	Serum concentration T_4 (μg/dl) (mean ± SD)	T_3 (ng/dl) (mean ± SD)	TSH[a] (μU/ml)	Urinary concentration SCN^- (μg/dl) (mean ± SD)	I^{-a} (mg/dl)	Urinary SCN^-/I^- (mean ± SD)
Boyabo	11	44 ± 13	9.4 ± 1.1	161 ± 46	4.32	0.93 ± 0.19	3.71	0.30 ± 0.20
Diobo	13	35 ± 14	9.9 ± 1.9	198 ± 49	4.8	1.15 ± 0.44	4.09	0.29 ± 0.15
Linzamba	24	28 ± 12	6.8 ± 1.5	206 ± 59	3.6	1.87 ± 0.38	4.95	0.42 ± 0.24
Lisombo	40	-	6.8 ± 1.8	136 ± 34	-	0.92 ± 0.54	1.94	0.47 ± 0.40
Likwangala	32	31 ± 13	5.8 ± 1.2	151 ± 31	3.6	0.71 ± 0.38	2.07	0.40 ± 0.27
Libobi	24	16 ± 7	8.1 ± 1.8	179 ± 29	2.5	1.67 ± 0.83	4.40	0.39 ± 0.23

[a]Measurement is the geometric mean.

Table 21. Coefficients of correlation between the epidemiological data and the measurements of thyroid function and goitrogenic factors.[a]

Indicator	Number of villages	Total goitre	Visible goitre	Cretinism
T_4	39	-0.67***	-0.75***	-0.44**
Log TSH	38	0.50***	0.65***	0.55***
Serum SCN^-	39	0.36*	0.31*	0.55***
Urinary SCN^-	50	0.25*	0.14NS	0.29*
Urinary I^-(log)	50	-0.75***	-0.68***	-0.48**

[a]Significance : NS = not significant; * P < 0.05; ** = P < 0.01; *** = P < 0.001.

(serum T_4 and TSH, urinary iodide, serum and urinary thiocyanate) calculated from the mean values in the test villages. There was a very significant correlation between prevalence of goitre and urinary iodide concentration ($r = 0.75$; $r^2 = 0.56$) as well as the indicators of thyroid function (T_4 and TSH). The positive correlation between prevalence rates and thiocyanate levels was less marked, a better correlation being obtained with serum than with urinary levels.

Table 22 shows the simple coefficients of correlation between the indicators of thyroid function (T_4, T_3, and TSH) and serum and urinary thiocyanate and urinary iodide. These coefficients were calculated from the individual values of inhabitants of the various villages. There were very significant correlations between urinary iodide and T_4 (positive) and TSH (negative). The percentages of explained variance, which correspond to r^2, were 21, 12, and 0.2 for T_4, TSH, and T_3 respectively. Serum T_3 did not seem to be related to urinary iodide but was related to thiocyanate levels and particularly serum thiocyanate. The correlations between thiocyanate and T_4 and TSH were also more evident when calculated from the serum thiocyanate values. The correlations were the inverse of those observed with urinary iodide.

To bring out antagonistic and combined effects of iodide and thiocyanate, we calculated the multifactorial relations linking the prevalence of goitre and biologic data to environmental factors.

Table 23 presents the equations of multiple regression with their multiple coefficient of correlation, the value of the F test of Fisher, and the associated probability.

Discussion

The epidemiological and biologic data from Ubangi confirmed the results of previous studies in other regions of endemic goitre (Chopra et al. 1975; Delange et al. 1976) as well as the preliminary results from Ubangi (Thilly et al. 1977). In contiguity with the description of the endemic region in the Uélés (De Visscher et al. 1961), we observed extreme goitre endemia in Ubangi with high prevalence of goitre and cretinism and an elevated frequency of abnormalities of thyroid function. As in the Uélés or in Kivu, the most frequently detected form of cretinism was myxedematous, with a prevalence 20 times that of neurologic cretinism. Up to 9% of the population in certain villages was affected.

The distribution of goitre as a function of age and sex was comparable to results in the literature (Delange 1974; Delange et al. 1976), confirming the existence of a particularly

Table 22. Coefficients of correlation between measurements of thyroid function and goitrogenic factors.[a]

Indicator	Serum SCN^- (sample size)	Urinary SCN^- (sample size)	Urinary I^-(log) (sample size)
T_4	-0.33*** (452)	-0.24*** (415)	0.46*** (282)
TSH (log)	0.24*** (468)	0.12** (391)	0.34*** (256)
T_3	0.16*** (469)	0.12** (432)	0.04NS (297)

[a]Significance: NS = not significant; * = P < 0.05; ** = P < 0.01; *** = P < 0.001.

Table 23. Correlations between the prevalence of goitre and the measurements of thyroid function and environmental factors.[a]

Parameter	Equation	r	F	df	Significance	r^2
Total goitre (TG)	$TG = 91.1 - 23.7\ UI^- + 6(S\text{-}SCN^-)^2$	0.79	19.3	2/23	$P < 0.001$	0.62
T_4 (μg/dl)	$T_4 = 4.9 + 2.6 \log UI^- - 2.6 \log S\text{-}SCN^-$	0.57	51	2/217	$P < 0.001$	0.32
TSH (μU/ml)	$\log TSH = 0.89 - 0.40 \log UI^- + 0.14 \log U\text{-}SCN^-$	0.37	19.4	2/246	$P < 0.001$	0.14
T_3 (ng/dl)	$T_3 = 182 - 34.5 \log S\text{-}SCN^-$	0.19	8.4	1/233	$P = 0.004$	0.04

[a]Urinary iodide: UI^- (μg/dl); serum thiocyanate = $S\text{-}SCN^-$(mg/dl); urinary thiocyanate = $U\text{-}SCN^-$ (mg/dl); multiple coefficient of correlation = r; Fisher test = F; degrees of freedom = df; significance of F test = P; percentage of explained variance = r^2.

critical period for adaptation of thyroid function. This period coincides with puberty in men and covers puberty and the childbearing years in women. The distribution of goitre resembles the evolution in urinary thiocyanate levels as a function of age and sex in the same region (Bourdoux et al. 1978).

Analysis of the geographic spread of endemic goitre revealed a homogeneous zone of severe goitre endemia in the north and centre of the subregion. In the south and west, along the Ubangi river, the prevalence values were clearly lower, but the boundaries of the severely affected zone did not coincide with any administrative, ethnic, or geologic division. Analysis by area provided a first impression that the endemia subsides as one approaches the river and swampy zone in the south; analysis by village confirmed it. This observation is similar to previous findings that in northeast Zaire, endemic goitre is not found along the rivers (De Smet 1960).

An examination of the geographic distribution of variations in thyroid function revealed the same marked distinction between a severely affected and a mildly affected zone. Markedly increased radioactive iodine uptake and serum TSH, moderately elevated T_3, and decreased T_4 were observed in the severely affected zone, whereas the values in the mildly affected zone were comparable to those in Belgium.

These differences combined with a severely reduced iodine supply and increased thiocyanate levels in the zone of severe endemic goitre. Iodine deficiency has a well known goitrogenic action (Studer et al. 1968; Koutras 1974; Thilly et al. 1972). Thiocyanate has been recognized as having an antithyroid action (Vanderlaan et al. 1947; Wollman 1962) and has been observed in other regions of endemic goitre in Kivu (Delange 1974), Europe (Silink 1964), and Nigeria (Ekpechi 1967).

The origin of the thiocyanate may be traced to the amounts of cassava ingested, cassava being one of the mainstays of the diet (Delange 1974; Bourdoux et al. 1978). The populations living along the rivers and in the swampy zones of south Ubangi have much lower serum and urinary thiocyanate levels, substituting fish for cassava in their diet (chapter 6) and, hence, perhaps also increasing iodine intake.

From a geologic point of view, the division of the subregion of Ubangi does not correspond with the limits of the markedly affected zone, although one cannot exclude the existence of aggravating or protecting factors in the soil as has been described elsewhere (Gaitan et al. 1974).

The study of environmental factors showed that the highest values for goitre and cretinism correspond to the association of severe iodine deficiency and the presence in the body of increased amounts of thiocyanate. Thiocyanate, the product of the catabolism of cyanide contained in food, significantly exacerbates the abnormalities of thyroid function induced by iodine deficiency. An increase in thyrotropic stimulation maintains thyroid function within the limits of normal in most cases but may be associated with an increase in thyroid volume, especially during the critical periods of life (growth, puberty, and maternity). In some instances the compensatory mechanisms are insufficient and hypothyroidism ensues, detectable biologically and sometimes clinically.

Chapter 5

Antithyroid Action of Cassava in Humans

P. BOURDOUX, F. DELANGE, M. GÉRARD, M. MAFUTA, A. HANSON, AND A.M. ERMANS

Experimental studies in animals have demonstrated a goitrogenic effect of cassava (Ekpechi et al. 1966; Ermans et al. 1972; Van der Velden et al. 1973), particularly in the rat. Also, in several regions where cassava is the mainstay of the inhabitants' diet, endemic goitre is associated with elevated serum and urinary thiocyanate (SCN) levels (Nwokolo et al. 1966; Ekpechi 1967; Oluwasanmi and Alli 1968; Delange and Ermans 1971a; Ekpechi 1973; Aquaron 1977). The studies carried out on Idjwi Island, Lake Kivu, Zaire, have emphasized the possible role of cassava in the etiology of endemic goitre (chapter 2). They suggested that eating cassava — the basic foodstuff of populations living in the tropics — increased the elimination of iodine in the urine and played a determining role in the etiology of endemic goitre in areas of limited iodine supply.

The intervention of goitrogenic foods is still controversial. Fifty years ago a goitrogenic effect associated with the intake of certain vegetables was first observed (Chesney et al. 1928). Later, the findings were confirmed for other foods (van Etten 1969). However, no definitive evidence of the role of such plants in the etiology of endemic goitre in human beings had been provided. The putative foods belonged, in general, to the family of Cruciferae and contained thioglucosides (van Etten 1969) whose catabolism produces substances such as thiocyanate (Vanderlaan and Vanderlaan 1947) or goitrine (Greer 1962). The antithyroid properties of these substances are well documented. In the last 20 years, other plants have been implicated as goitrogens (reviewed by Delange and Ermans 1976). They are capable of liberating hydrocyanic acid (cyanogenesis) when they are ingested due to degradation of their cyanogenic glucoside components (Montgomery 1969). The liberated cyanide is detoxified, converted to thiocyanate, by the body (chapter 1). Among the cyanogenic plants, cassava, in which the principal glucoside is linamarin (Clapp et al. 1966) is clearly the most important in terms of human intake. Specifically, cassava is the basis of the people's diet (Burssens 1958; chapter 6) in the region of Ubangi (northwest Zaire) that is affected by particularly severe endemic goitre. The epidemiological characteristics of this region are considered in chapter 4.

Evaluation of the degree of thiocyanate overload in the population of Ubangi and estimation of its repercussions on the capacity of the thyroid to adapt to iodine deficiency have clarified the role played by cassava in the etiology of endemic goitre.

Evaluation of Thiocyanate Levels in the Ubangi

Cassava is the basic foodstuff of the populations of Ubangi (chapter 6), and the biologic effects of this goitrogenic food are apparent. They are patently demonstrated in our studies of serum and urinary thiocyanate levels and daily urinary thiocyanate excretion values (Table 24). One group in our study was Ubangi adolescents from a rural environment who were away at boarding school (group I). They

Table 24. Comparison of serum and urinary thiocyanate (SCN^-) levels in Ubangi patients (groups I, II, and III) and in Belgian controls (group IV).[a]

Groups	Origin	Age (years)	Serum SCN^- (mg/100 ml) (mean ± SE)	Urinary SCN^- (mg/100 ml)	Urinary SCN^- (mg/d) (mean ± SE)
I	Ubangi: boarding school	9–16	0.26 ± 0.03 (22)	0.65 ± 0.03 (74)	4.92 ± 0.33 (74)
II	Ubangi: rural district	0–80	1.06 ± 0.07 (102)	1.92 ± 0.06 (386)	11.61 ± 0.92 (53)
III	Ubangi: rural district	10–30	0.81 ± 0.02 (989)	1.59 ± 0.04 (1065)	12.49 ± 1.09 (57)
IV	Belgium: controls	10–30	0.22 ± 0.02 (78)	0.60 ± 0.07 (23)	5.37 ± 1.07 (23)

[a]Figures in parentheses are the number of determinations.

had normal serum thiocyanate levels (0.26 mg/dl), which were almost identical to Belgian subjects taken as controls (group IV: 0.22 mg/dl). Daily thiocyanate excretion was also similar in the two groups, 4.92 and 5.37 mg/day respectively. In contrast, serum thiocyanate levels in rural populations subsisting almost exclusively on cassava were significantly increased (group II: 1.06 mg/dl; group III: 0.81 mg/dl; $P < 0.001$). An equally significant increase in urinary concentrations and daily excretion of thiocyanate was also evident. Even though there was an important dispersion in the serum thiocyanate levels in groups II and III, 30% of the subjects had levels between 1 and 2 mg/dl, and 7% had levels exceeding 2 mg/dl; i.e., 10 times the serum level in Belgian controls. Taken as a whole, the rural populations demonstrated a mean serum thiocyanate level 4–5 times higher than the Belgians; however, the thiocyanate levels in Ubangi adolescents at boarding school were normal, probably because their diet contained imported foods and, only occasionally, cassava.

The antithyroid properties of thiocyanate have been well documented in numerous studies (Vanderlaan and Vanderlaan 1947; Wollman 1962). The observation of elevated thiocyanate levels in a population may be considered to be a most valid indication of the consumption of a goitrogenic substance. Such observations have been made in studies of other regions of endemic goitre, notably in central Europe (Silink 1964), Kivu (Delange and Ermans 1971b), Nigeria (Nwokolo et al. 1966; Ekpechi 1967, 1973; Oluwasanmi and Alli 1968), and in Cameroon (Aquaron 1977). In animals, an increased excretion of thiocyanate was demonstrated to be associated with the ingestion of goitrogenic substances identified as thioglucosides (Langer 1964).

Thiocyanate Relationships

Relation between serum and urinary thiocyanate: It is important to determine whether serum or urinary thiocyanate is the more accurate indicator of thiocyanate overload. The relation between serum and urinary thiocyanate concentrations in Ubangi subjects is shown in Fig. 36. It is apparent that at levels higher than 0.5 mg/dl the serum concentration increases far less than the urinary concentration: one observes that for a 10-fold increase in urinary thiocyanate (0.5 to 5.0 mg/dl) the serum concentration only increases by a factor of three (0.5 to 1.5 mg/dl). This relation, detected in a large segment of the population, suggests that the measurement of thiocyanate in a random urine sample is a better indicator of thiocyanate overload than is a determination of serum thiocyanate. This is true despite possible errors introduced by extrapolating the results from a single urine collection.

Thiocyanate concentration as a function of age and sex: In the rural population of Ubangi (Table 24

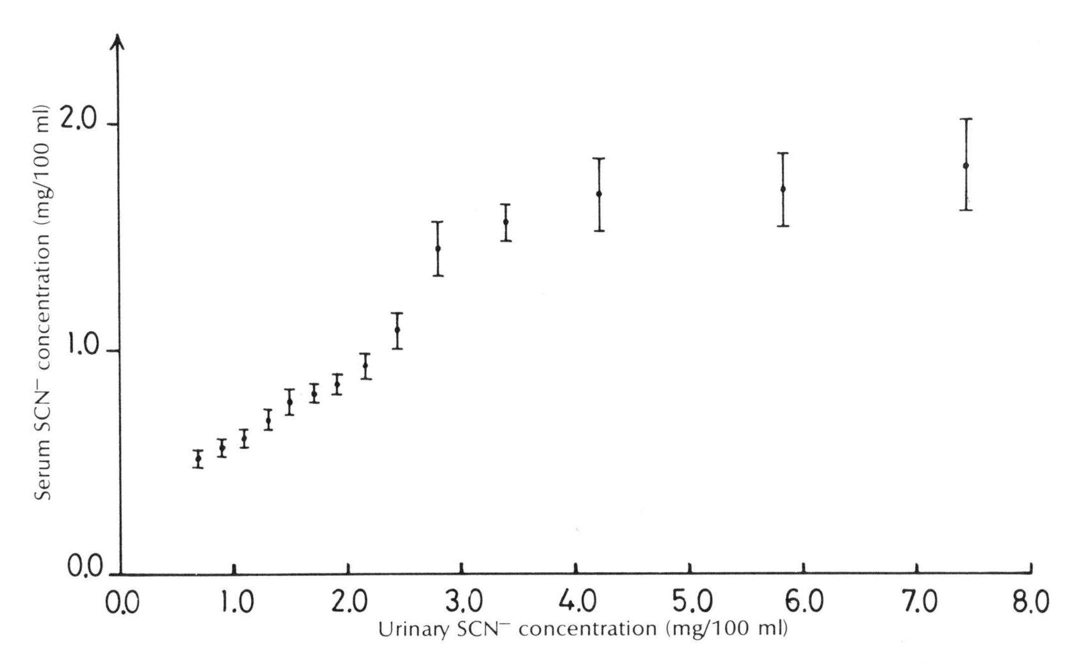

Fig. 36. Relationship between serum and urinary thiocyanate in 952 pairs of samples from groups I, II, and III (all Ubangi).

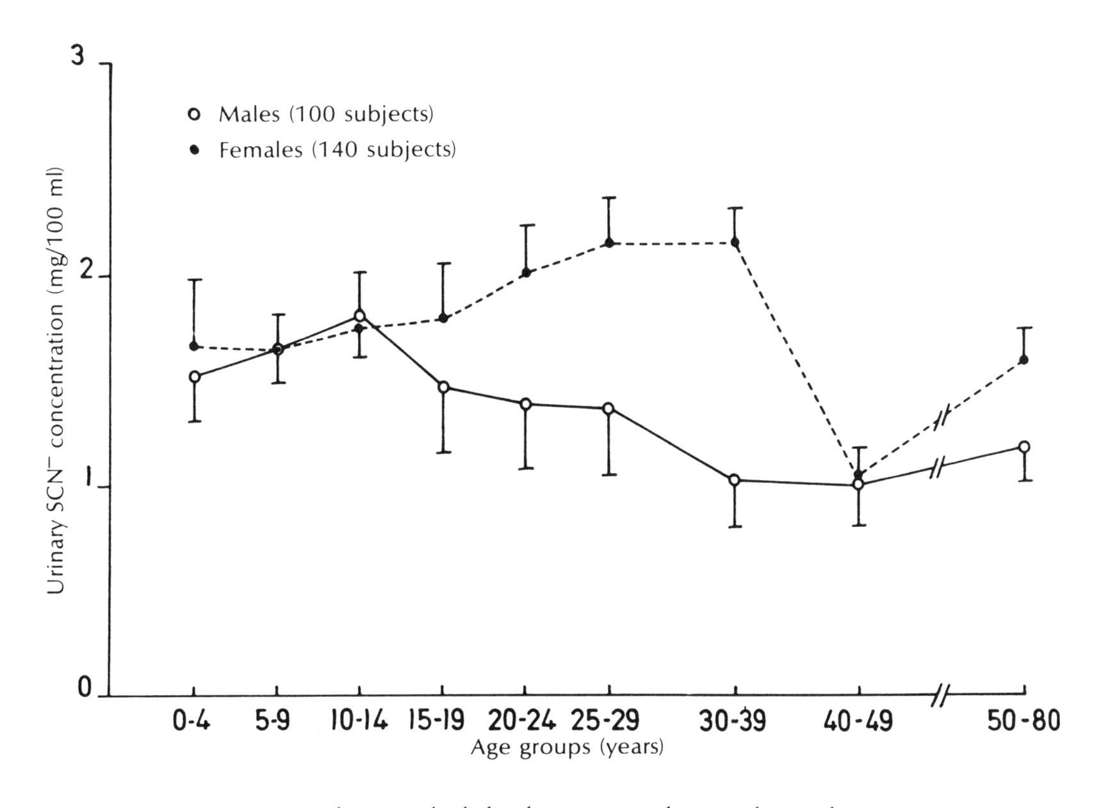

Fig. 37. Urinary thiocyanate level plotted against sex and age in Ubangi endemic goitre area.

groups II and III), the mean thiocyanate concentration in urine varied according to age and sex of person tested (Fig. 37). Up to approximately age 15, female and male subjects had similar values, but men and women, ages 25–40, had significantly different values. In women, urinary thiocyanate increased gradually to a maximum between 25 and 40 and then decreased. In men the urinary concentration decreased progressively from the value at age 15 to a minimum value between ages 40 and 49. After age 40–49 the values in the two sexes corresponded again. It is interesting to note that the variations in urinary thiocyanate concentration levels can practically be superimposed on the distribution of goitres in the Ubangi and Kivu endemic goitre areas (chapter 4; Delange and Ermans 1976).

Thiocyanate concentration as a function of size of goitre: In three separate villages, urinary thiocyanate concentration levels were proportional to size of goitre (Fig. 38, upper panel); in two of these villages (Bogene and Fulu), the differences between nongoitrous subjects and those with large goitres were statistically significant ($P < 0.025$). However, the analysis of urinary iodide concentration in two villages revealed no correlation with goitre size (Fig. 38, lower panel).

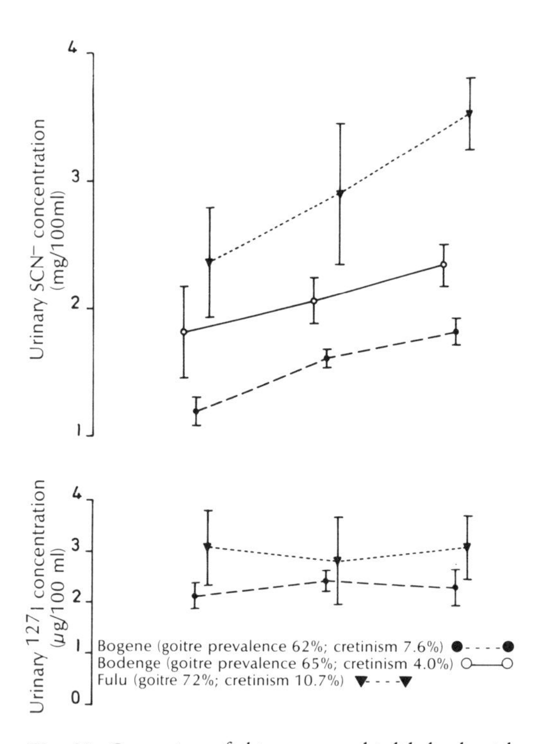

Fig. 38. Comparison of thiocyanate and iodide levels with thyroid gland size in Ubangi villages.

The relation between the degree of thyroid hyperplasia and urinary thiocyanate concentration was unexpected and difficult to explain. Although the concentrations in subjects without goitre differed from one village to the other, the direct correlation with goitre size was observed in all three villages. The urine collections were performed in the three villages at different times of the year. Thus, the different basal values could reflect the seasonal variations in cassava intake or cyanide content in the food consumed.

Effects on Metabolism

Effect of a large increase in cassava consumption on thiocyanate metabolism: For 3 days, inhabitants of Ubangi consumed large amounts of cassava or rice (controls). Two investigations were performed, one among adolescents at boarding school (group I) and the other in the rural environment (group II). The mean serum and urinary thiocyanate concentrations measured immediately before and after the high cassava or rice intake are shown in Table 25.

This massive ingestion of cassava roots provoked an increase in serum and urinary thiocyanate both in inhabitants of the traditional rural setting (group II) and youngsters at boarding school (group I). This increase was highly significant and more striking in the young women of group I who had lower prediet values. In these subjects, the increase in serum thiocyanate concentration was significant both for the mean value and the paired results. In the subjects with traditional food habits, the thiocyanate concentrations were elevated at the outset, and the change was less spectacular. This is probably because the supplementary increase in thiocyanate was blunted by a large preexistent thiocyanate pool. It is essential to note that 3 days of a cassava-free diet was sufficient to induce a significant decrease in serum and urinary thiocyanate levels in inhabitants of rural districts.

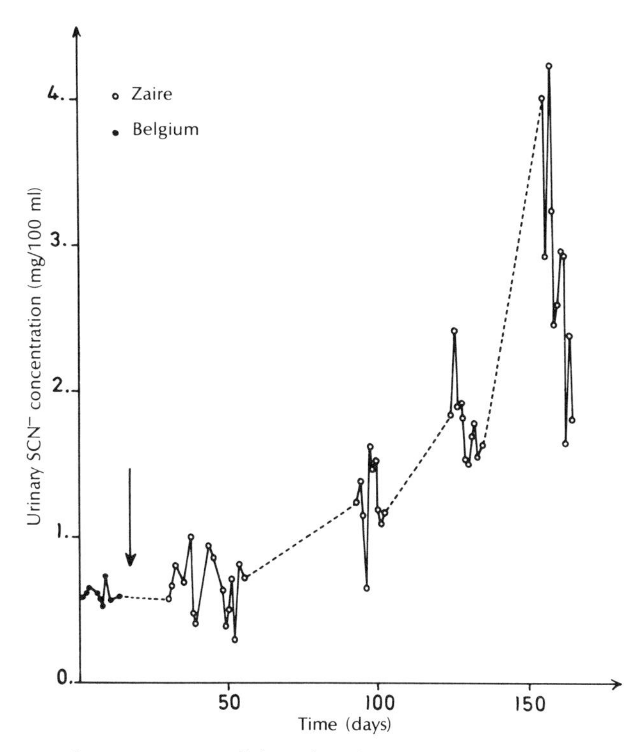

Fig. 39. Curve of urinary thiocyanate in one of the authors (M.M.) in Belgium (no cassava) and in Zaire (cassava).

If these observations do not exclude the possibility that other foods are involved, at least they indicate clearly that cassava plays the major role in the thiocyanate overload found in the Ubangi population.

Effect of chronic cassava intake on thiocyanate metabolism: The effect of chronic cassava intake was studied in an inhabitant of the Ubangi region of Zaire. After 5 months in Belgium, during which he did not eat cassava, he returned to Zaire and resumed traditional eating habits. Fig. 39 shows the values for urinary thiocyanate as a function of time. After 5 months without cassava, the mean value was 0.59 ± 0.02 (SE) mg/dl, identical to that observed in

Table 25. Changes in serum and urinary thiocyanate levels after ingestion of cassava or rice in two groups of Ubangi patients without (group I) or with (group II) regular cassava diet.

Groups	Parameters	Diets	Basal levels	Levels after change in diet	*t-test*[a]	Mean of individual changes (%)	Paired *t-test*[a]
I: boarding school	Serum SCN^-	Cassava	0.32	0.47	$P < 0.05$	+ 51	$P < 0.005$
(sample = 10)	(mg/100 ml)	Rice	0.21	0.18	NS	− 10	NS
	Urinary SCN^-	Cassava	0.71	0.88	NS	+ 33	$P < 0.005$
	(mg/100 ml)	Rice	0.58	0.56	NS	− 1	NS
II: rural district	Serum SCN^-	Cassava	1.05	1.02	NS	+ 3	NS
	(mg/100 ml)	Rice	0.82	0.64	NS	− 18	$P < 0.01$
	Urinary SCN^-	Cassava	2.08	2.33	NS	+ 20	$P < 0.05$
	(mg/100 ml)	Rice	1.73	1.54	NS	− 9	$P < 0.05$

[a]NS = not significant.

Belgian controls (Table 24). This value did not change (0.63 ± 0.05 mg/dl) during the first 7 weeks following his return to Zaire; however after 3 months, it increased strikingly to 1.23 ± 0.09 mg/dl and then to 3.06 ± 0.27 mg/dl 5 months after resumption of a cassava-based diet ($P < 0.001$). The results of this investigation suggest an important cumulative effect that could result from a relatively long biologic half-life of the thiocyanate ion (± 5 days — Ermans, A.M., unpublished data).

Effect of cassava intake on iodine metabolism: The effect of cassava intake on iodine metabolism was studied in adolescents at boarding school (Table 25, group I) fed cassava or rice during a 24-hour period. Two groups of 22 subjects were constituted such that the distribution of values of radioactive iodine uptake was equivalent in the two groups.

The subjects receiving cassava demonstrated a twofold increase in the excretion of stable iodine as compared with the subjects receiving rice: 18.1 ± 1.6 versus 9.3 ± 1.1 μg/24hours ($P < 0.001$) and a decreased 24-hour radioactive iodine uptake: 87.0 ± 2.0% versus 92.5 ± 1.5% ($P < 0.05$). The urinary thiocyanate excretion was similar in the two groups.

Thyroid Secretion and TSH Levels During Iodine Deficiency and Thiocyanate Overload

We also investigated the effects of iodine deficiency alone as opposed to those resulting from iodine deficiency plus thiocyanate overload (Table 26) by comparing thyroid function in two different groups. The boarding school group (group I) had a mean serum thiocyanate concentration of 0.26 mg/dl, a value comparable with that of Belgian controls, whereas the value for subjects living in a rural environment (group II) was 1.06 mg/dl. Notwithstanding this difference in serum thiocyanate levels, the daily stable iodine excretion was similar in groups I and II and markedly lower than the iodine excretion observed in Belgian controls ($P < 0.001$). This observation indicates a relatively constant degree of iodine deficiency in groups I and II.

However, the subjects in group II, with elevated serum thiocyanate concentrations, presented significant alterations of thyroid function:

- Decreased 24-hour thyroid ^{131}I uptake: 79.1 versus 89.0% ($P < 0.001$);
- Decreased serum T_4: 4.7 versus 7.3 μg/dl ($P < 0.01$); and
- Increased serum TSH: 31.8 versus 5.7 μU/ml ($P < 0.001$).

The combination of epidemiological and biologic data collected in Ubangi is characteristic of severe endemic goitre and is comparable with that reported for other regions of endemic goitre (Delange and Ermans 1976; Chopra et al. 1975).

The significant degree of iodine deficiency prevalent in Ubangi could explain the observed

Table 26. Parameters of iodine metabolism in Ubangi patients (groups 1 and II) and in Belgian controls (group III).[a]

Groups	Age (years)	Urinary $^{127}I^-$ (μg/d) (mean ± SE)	TSH (μU/ml) (mean ± SE)	T_4 (mean μg/100 ml ± SE)	24-h Thyroid uptake (% of dose) (mean ± SE)	Serum SCN^- (mg/100 ml) (mean ± SE)
I: Ubangi, boarding school	9–16	11.5 ± 0.9 (103)	5.7 ± 0.9 (22)	7.3 ± 0.5 (22)	89.0 ± 1.0 (70)	0.26 ± 0.03 (22)
II: Ubangi, rural district	9–16	13.5 ± 2.1 (35)	31.8 ± 3.2 (41)	4.7 ± 0.6 (41)	79.1 ± 1.0 (34)	1.06 ± 0.07 (41)
III: Belgium controls	10–30	51.2 ± 5.8 (38)	2.6 ± 0.3 (19)	8.4 ± 2.4 (36)	49.4 ± 2.3 (12)	0.22 ± 0.02 (75)

[a]Figures in parentheses are the number of determinations.

Table 27. Influence of thiocyanate injected intravenously into Belgian subjects.

SCN^- injected	Number of patients	24-h ^{131}I thyroid uptake (% of dose) (mean ± SE) Before injection	After injection	Serum SCN^- (mg/100 ml) (mean ± SE)	Uptake inhibition (%) (mean ± SE)
0 mg	10	–	–	0.50 ± 0.10	–
250 mg	6	41.9 ± 3.1	35.4 ± 2.7	1.41 ± 0.05	14 ± 3
500 mg	4	46.4 ± 1.8	29.9 ± 6.0	2.35 ± 0.26	34 ± 13

modifications in thyroid function, i.e., increase in thyroid ^{131}I uptake, decreased urinary iodine excretion, decreased serum T_4, and increased serum TSH. Such modifications and, in addition, thyroid hyperplasia can be produced experimentally in animals by the administration of iodine-deficient diets. Our observations demonstrated, however, that thiocyanate overload in conjunction with iodine deficiency induces more severe thyroid anomalies than those caused by iodine deficiency alone.

Effect of Thiocyanate on Radioactive Iodine Uptake under Conditions of Normal Iodine Supply

The effect of thiocyanate on thyroid radioactive iodine uptake was measured in 10 Belgian subjects with nontoxic goitre. Two uptake measurements were performed at an interval of 1 week, the second 24 hours after an intravenous injection of potassium thiocyanate. Serum thiocyanate concentrations were also measured. The results, including the percent inhibition of radioactive iodine uptake induced by thiocyanate administration are presented in Table 27. The initial uptakes were in the normal range for the Belgian population (30–55%). Thiocyanate administration induced a 14% inhibition in subjects with serum thiocyanate concentrations increased to 1–2 mg/dl and a 34% inhibition in subjects with serum thiocyanate concentrations between 2 and 3 mg/dl.

Conclusions

Cassava contains neither thioglucoside nor thiocyanate (Van der Velden et al. 1973). The thiocyanate detected is produced in the human body by detoxification of cyanide, which has been liberated from the cyanogenic glucoside, linamarin, contained in cassava. In the rat, a quantitative relation between the cyanide content of cassava roots and the amount of thiocyanate produced has been established (Osuntokun 1970). In the present study such a relation could not be demonstrated due to the enormous variability in the cyanide content of the food consumed (chapter 6). Nevertheless, taking the thiocyanate excretion of Belgian subjects as a reference point, one can estimate that the thiocyanate overload is 11 mg/day (Table 24). Assuming a quantitative conversion of cyanide to thiocyanate, a daily intake of 5–6 mg of cyanide is sufficient to account for this overload. The cyanide content of the food, measured in the form in which it is ingested (1.0–46.6 mg cyanide/kg fresh weight, chapter 6) not only equals but exceeds this level. In contrast to the data of Yamada et al. (1974), the chronic goitrogen overload is contained in a normal quantity of food and does not suppose an intake of several kilos per day.

The role of thiocyanate in the etiology of endemic goitre has already been proposed, as thiocyanate is known to accentuate the anomalies of iodine metabolism caused by iodine deficiency. A number of observations in our study supported this hypothesis and demonstrated that the association of endemic goitre and elevated thiocyanate levels in the observed population was highly significant:

- An elevated serum thiocyanate concentration ($>$ 1.0 mg/dl) was observed in one-third of the inhabitants of Ubangi. Such a concentration explains a significant inhibition of thyroid uptake under conditions of normal iodine supply. In addition, experimental studies (Maloof and Soodak 1964) have shown that thiocyanate uptake is accelerated in thyroid glands subjected to intense stimulation.

- A decreased thyroid radioactive iodine uptake and increased urinary iodine excretion was observed after substantial cassava intake. These observations confirm previous data of Delange and Ermans (1971b) for humans and those of Miller et al. (1975) for animals, which indicate that under acute conditions cassava ingestion perturbs the mechanism of adaptation of the thyroid to iodine deficiency.
- In populations with the same degree of iodine deficiency, thiocyanate overload induced more severe alterations of the parameters of thyroid function. In our study of iodine-deficient subjects, the group with elevated serum thiocyanate levels had lower serum T_4 and higher serum TSH levels than did the group with normal serum thiocyanate levels. This suggests that the intense stimulation of the iodide pump in the presence of elevated thiocyanate levels results not only from a process of adaptation to iodine deficiency but also from competition of the thiocyanate ion for the iodine-transport mechanism.

In summary, the iodine deficiency affecting Ubangi can explain the existence of endemic goitre in this region, characterized by classic alterations in iodine metabolism. However, this study emphasizes the role of thiocyanate overload whose effects are superimposed on those of iodine deficiency. The inhabitants of the region submitted to this double jeopardy can only maintain normal thyroid function at the price of extreme thyrotropic stimulation. It is logical to postulate that part of this stimulation exerts a compensatory effect on the competition between thiocyanate and iodide for the thyroid iodine-transport mechanism. Our observations confirm that the source of thiocyanate overload in Ubangi is closely related to cassava intake.

Chapter 6

Foods Consumed and Endemic Goitre in the Ubangi

E. SIMONS-GÉRARD, P. BOURDOUX, A. HANSON, M. MAFUTA, R. LAGASSE, L. RAMIOUL, AND F. DELANGE

The results described in chapters 4 and 5 establish that cassava has an antithyroid effect in humans and plays a role in the etiology of endemic goitre in Ubangi.

The purpose of the present work was to evaluate the relation between dietary habits of the populations of Ubangi–Mongala and epidemiological and metabolic characteristics of endemic goitre in this region. Particular attention was devoted to the role played by the consumption of cassava and the methods of preparing this basic food.

The study was divided into four distinct stages:

- Determination of the foods consumed and their methods of preparation;
- Investigation of food consumption as a function of the ethnic groups and the geographic location of the villages;
- Elucidation of the relationships between the epidemiological and metabolic characteristics of endemic goitre and nutritional habits; and
- Determination of the cyanide content of the foods.

Our nutritional survey was carried out in parallel with the program of treatment with iodized oil (chapter 7). We surveyed, systematically, the five zones of the subregion of Ubangi, i.e., Gemena, Libenge, Bosobolo, Budjala, and Kungu and collected some data from the subregion of Mongala.

We paid special attention to a north–south axis passing through Gemena and Budjala (Fig. 20). Epidemiological data had shown that goitre prevalence decreased along this line, thus providing a particularly favourable situation for us to assess the role of the nutritional habits in the development of goitre.

Methods

Determination of the foods consumed and their methods of preparation: Our team questioned local medical unit personnel, village notables, missionaries working in the villages, and the villagers themselves to determine what foods were consumed and how they were prepared. There were no discrepancies in the information collected from the different sources, and the data were confirmed by numerous observations in the field and in the habitations during preparation of the meals.

Investigation of food consumption: The most precise method for us to have measured food consumption would have been to monitor food consumed during a period of at least 1 year, noting seasonal variation (Vis et al. 1969, 1972). To follow this method, we would have had to live near the families and to weigh each food consumed during each meal. We did not have the means to implement such rigorous methods, so we based our qualitative and cross-sectional analysis on information gathered by means of food questionnaires (Gordon 1963; Jeliffe 1969; Lambrechts and Bernier 1961). The frequency and combinations of food

Sous-région

Date

Coll. loc

Village

sexe
âge
travail du chef de famille
travail de la femme/mère

fuku – ka
chickwangue – kwanga
bui (bouillie) – ko
mpondu – sabunda
nsongo (manioc) – tolo kadanga
moteke
mafuku (pain d'arbre)
mabenge (patate douce) – goleya
mayika (taro) – toko
ngondo – ngbese
masangu (maïs) – koni
loso (riz) – silifo
autres céréales
makemba – bo
soya
nguba (arachides) – nzo
bisapa (ignames) – gbasi
liboke (courge) – sa
feke feke (okra) – ngbali
makembo (champignon) – 'bua
feuilles vertes
autres légumes
nyama (viande) – sade
soso (volaille) – kola
likei (oeuf) – kulu
mbisi (poisson) – koyo
poisson séché
poisson en boîte
ndonge (termites) – dole
kalo (fourmis) – tinda
mbinzu (chenilles) – doko
mbembe (escargots) – ndaba
mabele (lait) – nu
mbila (noix de palme) – bete
mbuma (fruits) – walate
koko (canne à sucre) – kumba
masanga (vin de palme) – dô
ngako, rak (alcool) – dôwè
autre boisson
divers

Fig. 40. *Questionnaire used in food survey.*

consumed at each meal during a day were determined (Fig. 40).

The first section of the questionnaire identified the village and provided general information about the subject: sex, age, work performed by the head of the family, and the person questioned (that of the mother for children under 15). The second section listed 40 foods — in Lingala, French, and the local dialect — that might be consumed by the population, i.e., diverse preparations of cassava, cereals, vegetables, meats, poultry, fish, other animal proteins, fruits, and liquids.

Each person questioned was asked to enumerate the foods eaten during the 24 hours preceding the interview, delineating foods taken at and between meals and, in particular, raw cassava. These foods were entered on the questionnaire, divided for each subject into three columns corresponding to the three daily meals plus in-between meal consumption.

This method of recording facilitated the calculation of the frequency of consuming each food for a group of subjects examined in the same village and the average daily consumption of each food by each individual (horizontal analysis) as well as the determination of the composition of the meals (vertical analysis). The vertical analysis was only used in 13 villages where biochemical data on thyroid function and serum and urinary levels of thiocyanate were available to elucidate relationships between the nutritional information and goitre prevalence.

The nutritional questionnaire was administered in 42 villages during the mass treatment campaign. The geographic distribution of the villages is given in Fig. 41. We selected 1465 subjects at random (Reh 1963) by questioning 1 of every 10 villagers in line to receive an injection of Lipiodol; 46.4% were younger than 18, 47.5% were male, and 52.5% female. This distribution corresponds to that of the general population of the region investigated (Boute 1973).

The results obtained were coded for computer analysis.

Determination of the cyanide content of the foods: We collected foodstuffs from 50 different villages scattered throughout Ubangi including the 13 villages investigated in detail. The villagers were asked to supply fresh cassava roots identified by them as belonging to the sweet and bitter varieties, as well as foods to be prepared and foods ready to eat. As described in chapter 1, there were no consistent morphological features we could distinguish in the sweet and bitter varieties of fresh cassava roots. We limited our analyses of foods ready to eat and to be prepared to those available in villages near enough to our laboratory to allow rapid transport. For this reason, the number of foods analyzed for the "Gens d'eau," for example, was very limited.

Cyanide content was measured as described in chapter 1.

Results

Cassava and its derivatives consumed: The sweet variety of cassava roots is either eaten raw, mainly in-between meals, or boiled. One of the ways bitter cassava is used is *fuku*, a gruel of cassava flour to which variable amounts of corn are incorporated according to the season. The cassava is not soaked. The roots are peeled, cut into pieces, and dried in the sun for 1 or 2 days. Then they are bruised in a mortar with corn (Fig. 42) that has been steeped for 12–24 hours in water. The flour thus obtained is then grilled on a plank to halt the fermentation initiated by steeping the corn. The flour is eaten as a gruel prepared in boiling water.

Mpondu is the vegetable most frequently eaten by all the ethnic groups in our study. It is obtained from cassava leaves that are ground and then extensively boiled. Palm oil and sometimes ground peanuts are added.

Chickwangue, like *fuku*, is prepared from bitter cassava. The roots are soaked for 2–6 days, then mashed into a purée that is simmered to form a paste of firm and elastic consistency. The paste is enveloped in a palm or banana leaf.

Ntuka is a preparation similar to *chickwangue*, made from bitter cassava soaked for 3–5 days, then peeled, and steamed in a pot. Sticks are put in the bottom of the pot to support the cassava over the boiling water, which in turn is covered with leaves.

Moteke is a gruel obtained from the flour of bitter cassava. The cassava is first soaked in stagnant water for 1–2 days and then worked into a paste. The paste is divided into pieces that are dried in the sun. Then the pieces are ground into flour, which is boiled in water.

Proteins consumed: Fish is rarely eaten fresh except by the "Gens d'eau." More commonly, it is dried and smoked to ensure its preservation.

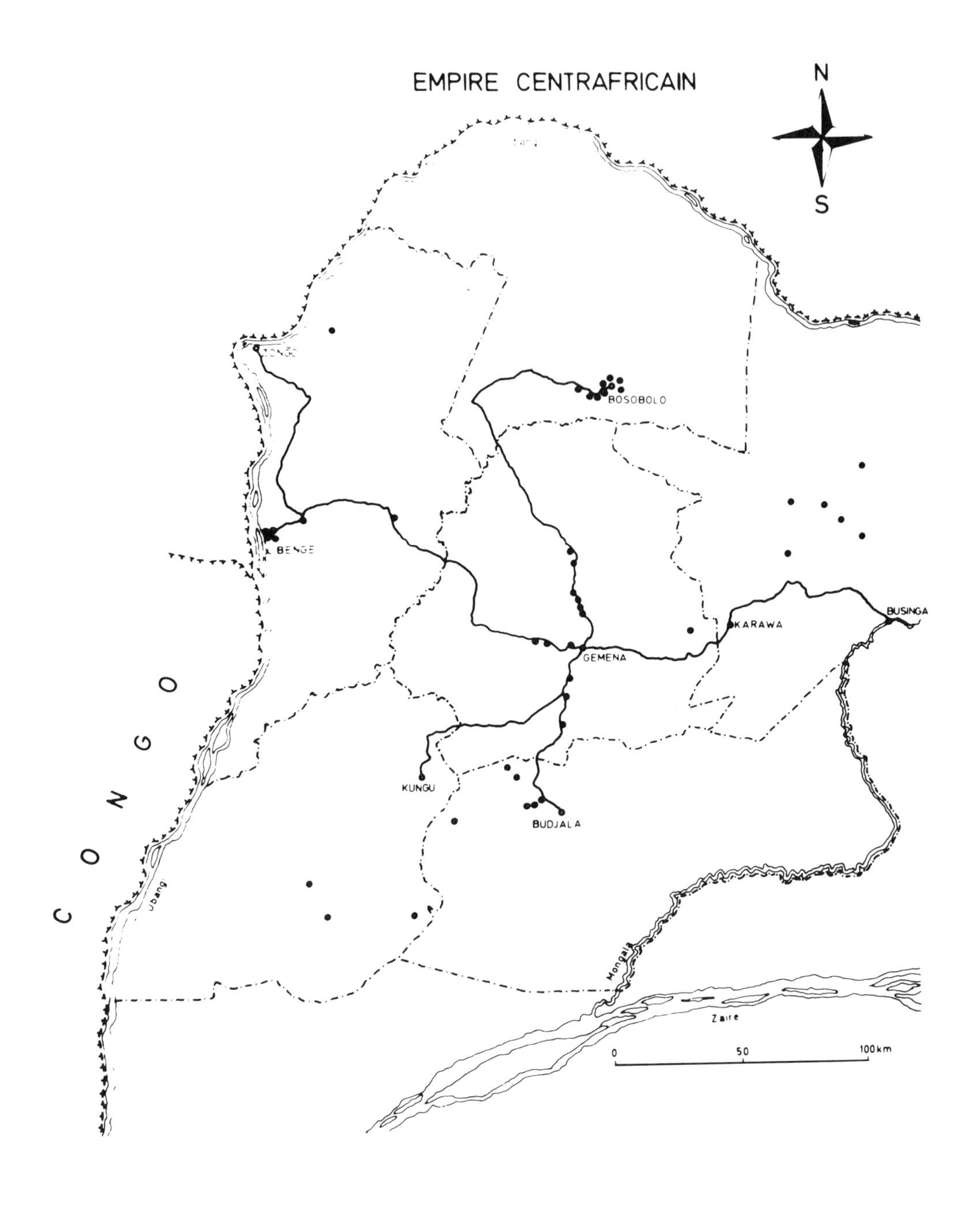

Fig. 41. Geographic distribution of the villages visited during the nutritional survey.

Fig. 42. *Preparation of fuku: (a) fresh roots and leaves of bitter cassava; (b) roots are peeled and cut into pieces; (c) pieces of cassava are dried in the sun; and (d) dried cassava pieces and corn are bruised in a mortar; the flour is scaved.*

Meat comes from hunting (monkey, antelope) and is fresh or cured. Goats, pigs, and poultry (chickens, ducks) have commercial exchange value and are rarely eaten.

Other foods consumed: Other than cassava leaves *(mpondu)*, which are the basic vegetable, occasionally one finds squashes, okra, mushrooms, and various green leafy vegetables. However these only form a negligible part of the diet.

Makemba is the Lingala term designating the plantain banana. This is consumed as a purée after having been boiled or as dumplings fried in palm oil.

The principal carbohydrates are sweet potatoes, yams, and taros. They are boiled and accompanied by vegetables and/or meat or fish.

Peanuts are eaten fresh or roasted. They are sometimes incorporated into other recipes *(mpondu)*.

Palm nuts are bruised in a mortar and boiled in water. When the mixture has cooled, the floating oil is carefully collected.

Frequency of foods consumed and daily average consumption: Table 28 shows the frequency of consumption of the principal foods in the six ethnic groups of the region: Ngbaka, Mbanza, Ngombe, Ngbaka–Mabo, Mongwandi, and "Gens d'eau." For the first two, we collected data in three different zones to determine whether the nutritional habits of the same ethnic group were influenced by the geographic dispersion of the villages. The table also presents results obtained in two cities: Libenge (in Ngbaka) and Bosobolo (in Ngombe), elucidating eventual modifications in diet in an untraditional setting. The results have been expressed as a percentage of the people questioned who had consumed a given food during the 24 hours preceding the interview.

The basic food of the Ngbaka, of the zones of Gemena, of Businga (subregion of Mongala), and of Libenge is *fuku*, which is eaten by 91, 92, and 87% of the subjects in the three zones respectively and is usually accompanied by *mpondu*. Less frequently it is associated with other vegetables or meat (3–15%), fresh or dried fish (1–21%), chickens (2–3%), or caterpillars, snails, and termites in season (from 4–37%). Other sources of carbohydrates are *makemba* (2–12%), *chickwangue* (0–14%), sweet cassava (4–12%), and corn in season (18–45%). Yams, taros, and sweet potatoes are not shown

Table 28. Percentages of people questioned (1465) who had consumed the principal foods within previous 24 hours.

Food	Ngbaka			Mbanza			Ngombe	Ngbaka–Mabo	Mong-wandi	"Gens d'eau"	Libenge (city)	Bosobolo (city)
	Gemena	Mongala	Libenge	Ghado	Boyado	Budjala	Bosobolo	Libenge	Budjala			
Fuku	91	92	87	89	77	57	17	63	1	–	38	17
Mpondu	69	51	68	54	87	69	68	88	32	28	67	54
Moteke	–	1	–	–	2	32	16	23	4	–	11	35
Chick-wangue	5	14	–	18	13	11	2	1	53	45	19	10
Sweet cassava	11	12	4	14	34	16	33	19	20	6	15	23
Corn	18	22	45	7	2	21	6	5	19	–	3	5
Ntuka	–	–	–	–	–	–	–	–	–	44	–	–
Rice	2	3.5	–	–	–	2	57	–	14	4	3	46
Makemba	12	5	2	7	28	24	18	11	33	45	43	19
Peanuts	29	22	11	14	26	41	40	1	27	15	18	40
Fresh fish	1	11	–	4	11	16	14	5	11	78	20	17
Dried fish	3	11	21	–	11	1	5	–	1	–	10	17
Meat	3	9	15	4	11	13	16	5	13	15	18	35
Poultry	2	3	–	–	2	2	2	1	1	–	1	4
Other animal pro-teins	13	37	4	64	–	30	–	–	63	–	5	1

Table 29. Average number of times/day that 12 foods were consumed by different ethnic groups and regions.

	Ngbaka			Mbanza			Mong-wandi	Ngombe	Ngbaka-Mabo	"Gens	Libenge	Bosobolo
Foods	Gemena	Mongala	Libenge	Ghado	Boyado	Budjala	Budjala	Bosobolo	Taba	d'eau"	(city)	(city)
Fuku	1.7	1.5	2.2	1.6	1.3	1.1	0.01	0.2	1.2	–	0.6	0.3
Mpondu	1.4	0.6	1.4	0.7	1.7	1.3	0.5	1.2	1.7	0.4	1.2	0.8
Moteke	–	–	–	–	0.02	0.6	0.05	0.2	0.5	–	0.2	0.5
Chick-wangue	0.05	0.2	–	0.3	0.1	0.2	1.00	0.03	0.01	0.8	0.3	0.1
Sweet cassava	0.2	0.2	0.2	0.2	0.4	0.2	0.3	0.4	0.2	0.1	0.2	0.3
Corn	0.03	0.20	0.5	0.1	0.04	0.3	0.3	0.1	0.05		0.03	0.1
Ntuka	–	–	–	–	–	–	–	–	–	1	–	–
Rice	–	–	–	–	–	–	0.2	0.9	–	–	0.03	0.7
Makemba	0.1	0.1	0.02	0.1	0.4	0.3	0.6	0.2	0.1	0.6	0.6	0.3
Fish	0.1	0.2	–	0.04	0.2	0.2	0.2	0.2	0.1	1.4	0.2	0.2
Dried fish	–	0.1	0.4	–	0.2	0.02	0.01	0.1	–	–	0.2	0.2
Meat	0.01	0.1	0.3	0.1	0.2	0.2	0.2	0.2	0.1	0.2	0.3	0.5

in Table 28. They are occasional replacements for *fuku*.

The Mbanza of the zones of Libenge, of Budjala, and of Mobayi Mbongo have very similar nutritional habits to those of the Ngbaka. Of those questioned, 67% had consumed *fuku* and 70%, *mpondu* within 24 hours. However, they ingest more *moteke* (up to 32%), *chickwangue* (11-18%), and *makemba* (7-28%).

The results obtained for the same ethnic groups in different localities were very similar, showing that the dietary habits generated by the ethnic group are not appreciably altered by geographic dispersion.

The Ngombe of the zone of Bosobolo subsist principally on rice (57%) and *mpondu* (68%). The rice can be replaced by boiled sweet cassava (33%), *moteke* (16%), *fuku* (17%), and *makemba* (18%). Meat and fish consumption is the same as in the Ngbaka and Mbanza.

The Ngbaka-Mabo of the village of Taba (north Libenge) ingest similar foods to those of the Ngbaka and Mbanza. The basic foodstuff is *fuku* (63%) and *mpondu* (88%). *Moteke* (23%) and sweet cassava are more important than in the preceding groups. The consumption of animal protein is low (5%).

The Mongwandi of Budjala have a diet composed of *chickwangue* (53%), *makemba* (33%), and *mpondu* (32%). Other sources of carbohydrate are sweet cassava (20%), corn at the time of harvest (19%), and rice (14%). Animal proteins are provided by fish and meat in proportions similar to those for the Ngbaka and Mbanza. In season, caterpillars and snails are often eaten (63%).

The "Gens d'eau," in contrast to the land dwellers, consume important amounts of fresh fish, nearly daily (78%). The fish is accompanied by *chickwangue* (45%) and/or *ntuka* (44%), *makemba* (45%), and *mpondu* (28%). A secondary source of animal protein is meat (15%). This group does not eat *fuku*.

The situation in urban centres as compared with that in the same ethnic groups in the traditional situation is characterized by:

- Decreased consumption of the traditional basic food (*fuku* for the Ngbaka, rice for the Ngombe);
- Increased consumption of *moteke* and *chickwangue*; and
- Increased consumption of animal proteins.

Average daily consumption of different foods: The values for average daily consumption of foods are indicated in Table 29. The Ngbaka, Mbanza, and Ngbaka-Mabo consume *fuku* 1.2–2.2 times per day and *mpondu* 0.6–1.7 times, whereas *chickwangue* is eaten 0.05–0.2 times per day. For the Mongwandi, *fuku* and *mpondu* are taken 0.01 and 0.5 times respectively, *chickwangue* 1.0 times. *Fuku* is not eaten by the "Gens d'eau." Their average consumption of *chickwangue* is 0.8 and of *ntuka* 1.0.

Composition of the meals: The analysis of the composition of the meals in the 13 villages studied is shown in Table 30. Because of the particular interest in the relation between cassava intake and endemic goitre, the results presented comprise essentially the different forms of cassava.

Villages 1–8 correspond to Ngbaka, Ngbaka–Mabo, and Mbanza (group I), villages 9–10 to Mongwandi (group II), and villages 11–13 to the "Gens d'eau" (group III). In all villages 67–97% of the inhabitants had consumed cassava in the 24 hours preceding the interview. The essential result illustrated in this table is that the ways of preparing cassava and the associations of the type of cassava ingested are very similar in the different villages of an ethnic group but very different from one group to another. In the villages of group I, cassava is essentially consumed as *fuku* and *mpondu*, and for one of the villages *moteke* in addition. In contrast, *fuku* is not found in villages 9–13 where in its place cassava is consumed essentially in *chickwangue* and *mpondu* in the villages of group II and *chickwangue* and *ntuka* in the villages of group III. The Mongwandi and the "Gens d'eau" also demonstrated an important consumption of *makemba*. As already mentioned the "Gens d'eau" have the highest intake of proteins.

Table 31 summarizes the characteristics of cassava consumption as a function of ethnic groups. The results were obtained by regrouping the data from the different villages in each of the groups I, II, and III. It is clear that although all the groups subsist on cassava, varying in frequency from 75–90%, there is an important distinction between those who consume *fuku* (group I) and those who consume *chickwangue* (groups II and III).

Relationships between the epidemiological and metabolic characteristics of endemic goitre and nutritional habits: Table 32 correlates the results obtained for the epidemiological and metabolic characteristics of endemic goitre and the serum and urine levels of thiocyanate (SCN) in the three groups. In passing from groups I to III one observes an important decrease in the prevalence of goitre and cretinism, a decrease in ^{131}I thyroid uptake and serum TSH, and a progressive increase in serum T_4.

Table 30. Percentage of meals containing cassava and its derivatives, banana (*makemba*), and proteins in 13 villages of Ngbaka, Ngbaka–Mabo, and Mbanza (group I), Mongwandi (group II), and "Gens d'eau" (group III).

Ethnic group	Village	Cassava	*Fuku*	*Moteke*	*Chickwangue*	*Ntuka*	*Mpondu*	*Fuku & mpondu*	*Makemba*	Animal proteins
Group I										
Ngbaka	1	93	82	–	2	–	39	39	2	15
Ngbaka	2	92	76	–	1	–	78	66	4	3
Ngbaka	3	86	80	–	–	–	49	46	2	25
Ngbaka–Mabo	4	90	69	–	5	–	39	32	–	34
Ngbaka–Mabo	5	96	58	23	1	–	78	50	6	10
Mbanza	6	97	72	–	6	–	55	45	3	8
Mbanza	7	87	58	7	7	–	53	34	5	23
Mbanza	8	91	61	1	6	–	75	54	18	21
Group II										
Mongwandi	9	78	–	3	55	–	22	–	16	15
Mongwandi	10	67	–	–	6	–	63	–	57	37
Group III										
"Gens d'eau"	11	69	–	–	38	21	12	–	32	52
"Gens d'eau"	12	88	–	–	9	79	8	–	14	81
"Gens d'eau"	13	82	–	–	38	33	25	–	21	68

Table 31. Percentage of meals consumed that contain cassava and its derivatives.

Group	Ethnic group	Cassava	*Fuku*	*Mpondu*	*Fuku & mpondu*	*Chickwangue*	*Ntuka*	*Moteke*	*Chickwangue, ntuka, & moteke*
I	Ngbaka	90	77	51	46	2	0	0	2
	Ngbaka-Mabo	96	58	78	50	1	0	23	24
	Mbanza	89	64	61	44	7	0	4	11
II	Mongwandi	75	0	43	0	43	0	3	46
III	"Gens d'eau"	78	0	15	0	30	41	0	71

The serum and urine values of SCN were strikingly elevated for the three groups. They were 1.5–2.0 times higher in the first group than in the two others. The ratio of urinary SCN to urinary iodide passed from 2.1 in group I to 0.75 in group II and to 0.39 in group III. The serum and urine concentrations of SCN in group I were significantly higher than those for groups II and III taken together (1.24 vs 0.65 mg/dl and 2.57 vs 1.49 mg/dl respectively, $P < 0.001$). Similarly the urinary SCN/I ratio was 2.10 ± 0.21 in group I and only 0.53 ± 0.04 for groups II and III ($P < 0.001$). Urinary iodide in group II was slightly but not significantly higher than in group I (ratio 1.3, $P > 0.05$), and in group III, it was very clearly higher (ratio 2.6, $P < 0.001$).

Table 33 shows the relationships existing between, on the one hand, the overall prevalence of goitre, visible goitre, and cretinism, and, on the other hand, the consumption of cassava in its various forms and of animal proteins. There was a direct though only weakly significant correlation between the three epidemiological variables and the consumption of cassava. This correlation becomes highly significant when one considers separately the consumption of *fuku* and *fuku-mpondu* ($P < 0.001$). There was an inverse correlation between the epidemiological characteristics and the consumption of *chickwangue* and of animal proteins ($P < 0.01$).

The relation between the prevalence of goitre and the consumption of *fuku* must be interpreted as a function of the concomitant modifications of iodine supply. The respective roles played by iodine and SCN supplies are illustrated in Fig. 43. This figure shows the correlations between the prevalence of goitre and urinary iodide concentration (upper panel) and urinary SCN/I ratio (lower panel). There was a significant inverse correlation ($y = 106.9\ e^{-0.3400}$, $r = 0.741$, $P < 0.01$) between goitre prevalence and urinary iodide concentration. Although this result confirms a classic concept, it is nevertheless striking to realize that the prevalence of goitre is approximately 20 to 30% for urinary iodide concentrations of the order of 4 μg/dl, i.e., the urinary concentration observed in Brussels where goitre is detected in 3% of the population. Goitre prevalence was significantly correlated with the urinary SCN/I ratio ($y = 46.5 + 23.9 \ln x$, $r = 0.799$, $P < 0.001$), and this ratio is independent of the absolute values obtained for each of the two variables.

Cyanide content of foods: Fig. 44 compares the

Table 32. Epidemiological and metabolic indicators for three groups investigated.[a]

Group	Prevalence of goitre (%) (mean± SE)	Prevalence of cretinism (%) (mean± SE)	6-h ^{131}I thyr. uptake (%) (mean± SE)	Serum levels: T_4 (μg/dl) (mean± SE)	Serum levels: TSH (μU/ml) (mean± SE)	Serum levels: SCN (mg/dl) (mean± SE)	Urinary levels: SCN (mg/dl) (mean± SE)	Urinary levels: ^{127}I (μg/dl) (mean± SE)	Urinary SCN/I ratio (mg/μg) (mean± SE)
I	60± 4	6.1± 1.0	55± 2	4.4± 0.1	27.0± 4.0	1.24± 0.03	2.57± 0.10	1.9± 0.1	2.10± 0.21
	(8)	(8)	(142)	(408)	(409)	(406)	(379)	(254)	(250)
II	34	1.1	42± 3	6.4± 0.4	3.8± 0.6	0.42± 0.04	1.32± 0.11	2.5± 0.3	0.75± 0.08
	(2)	(2)	(46)	(43)	(44)	(44)	(49)	(49)	(49)
III	17± 5	0	23± 1	6.8± 0.2	3.3± 0.5	0.78± 0.04	1.60± 0.07	4.9± 0.3	0.39± 0.02
	(3)	(3)	(75)	(77)	(76)	(76)	(77)	(77)	(77)

[a]Figures in parentheses are the numbers of villages or of subjects.

Table 33. Correlation coefficients between epidemiological indicators and consumption of cassava, its derivatives, and animal proteins in 13 villages.[a]

Epidemiological characteristics	Cassava	*Fuku*	*Mpondu*	*Fuku* & *mpondu*	*Chickwangue*	Fish & meat
Total prevalence of goitre (%)	0.63*	0.86***	0.74**	0.81***	-0.77**	-0.68**
Prevalence of visible goitre (%)	0.62*	0.83***	0.58**	0.75**	-0.60**	-0.78**
Prevalence of cretinism (%)	0.68*	0.77**	0.60**	0.75**	-0.60**	-0.70**

[a]Significance: *= P < 0.05; * = P < 0.01; *** = P < 0.001.

cyanide content of the roots of fresh cassava identified by the villagers as being of the sweet or bitter varieties. For both varieties there was an important variation between individual values, the concentrations measured in the bitter cassava samples ranging from 5 to 166 mg/kg and no results for sweet cassava falling below the minimum values for the bitter variety. Nevertheless, the mean value for sweet cassava [32.9 ± 2.9 mg (SE) HCN per kg] was approximately half that of bitter cassava

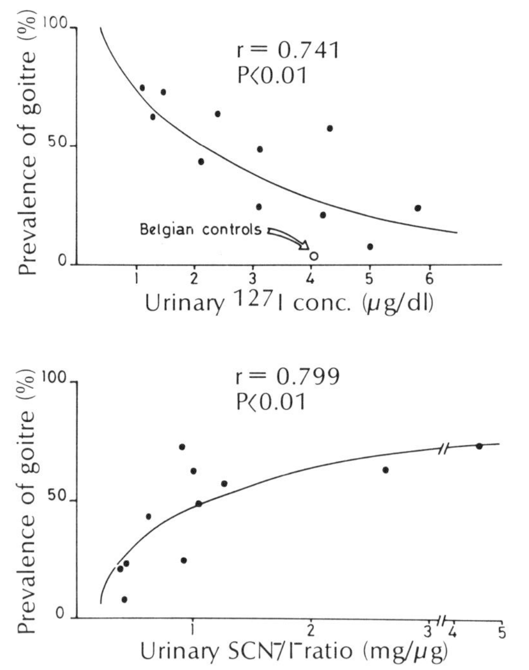

Fig. 43. Correlation between the prevalence of goitre and urinary iodide concentration (upper panel) and urinary SCN/I ratio (lower panel).

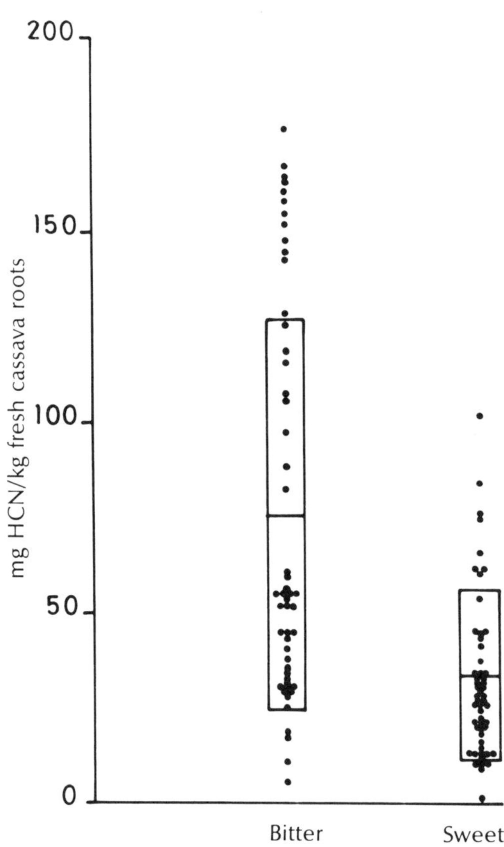

Fig. 44. Comparison of the cyanide content of fresh roots of sweet and bitter cassava. Individual values, mean ± SD.

(74.5 ± 7.0 mg/kg $P < 0.001$). In addition, in a given village, the roots presented as sweet systematically contained less cyanide than roots from bitter cassava.

Table 34 shows the cyanide concentrations in foods ready to eat and foods to be prepared before being cooked. In the first category the highest values were obtained for *fuku*, soaked cassava, *mpondu*, and cooked cassava leaves. The cyanide content of *chickwangue* is 2–4 times as low. Among the foods to be prepared, the highest level was found in fresh cassava leaves (92 mg/kg). The mixture of cassava and ground corn still contained approximately 20 mg HCN/kg. No appreciable quantities of cyanide were found in other foods not shown in the table with the exception of soybean, which contained a substantial concentration of the order of 17–47 mg/kg.

Discussion

This study showed that the customary diet of the populations of Ubangi is relatively varied, rich in carbohydrates, lipids, and proteins of vegetable and animal origin, particularly in the case of the "Gens d'eau." This explains the lack of overt problems of protein-calorie malnutrition in the region. These findings confirm data reported by Bervoets and Lassance in 1959.

Cassava is the basic foodstuff: 67–97% of the subjects questioned ate cassava in the 24 hours preceding the interview.

The essential contribution of this work is the demonstration that the methods of preparing cassava vary widely from one ethnic group to another and that these differences probably play a critical role in the development of goitre.

In the groups where the prevalence of goitre was the highest (Ngbaka, Ngbaka–Mabo, and Mbanza), cassava was consumed primarily as *fuku*. This food is prepared from bitter cassava, which has not been soaked and which contains important quantities of cyanide. In contrast, in the groups with a lower prevalence of goitre, cassava was eaten essentially in the form of *chickwangue* whose cyanide content is approximately four times as low as that for *fuku*.

Table 34. Cyanide content of foods ready to eat or to be prepared.

Food	Number of samples	HCN (mg/kg) (mean ± SE)
Ready to eat		
fuku	172	14.2 ± 0.7
soaked cassava	5	11.8 ± 4.4
mpondu	9	8.5 ± 0.7
cooked cassava leaves	4	7.3 ± 4.7
chickwangue	17	3.5 ± 0.4
cane sugar	8	2.6 ± 0.1
To be prepared		
cassava leaves	5	92.1 ± 21.6
cassava & corn	23	19.8 ± 2.5
cassava flour	4	4.5 ± 2.7
wild yams	5	7.3 ± 0.4
cultivated yams	4	8.5 ± 0.3
plantain banana (*makemba*)	8	4.4 ± 0.7

The cyanide content of the fresh roots from which these two foods are prepared was equally elevated. These findings account for the fact that the serum and urine concentrations of thiocyanate were increased for the entire population but clearly higher for the groups consuming *fuku*.

The demonstration of the direct role in the development of goitre played by the derivatives of cassava that are insufficiently detoxified was complicated by the finding in some villages of decreased iodine supply as well as thiocyanate overload. Our results show that the prevalence of goitre is critically linked to the balance between iodine supply and thiocyanate, independently of the absolute quantity of iodine available (Fig. 43, lower panel). They confirm the data reported in chapter 5 showing that, at a given level of iodine supply, the addition of thiocyanate gives rise to thyroid abnormalities characteristic of endemic goitre. Comparison between groups I and II (Table 32) shows that for a urinary iodide concentration slightly, but not significantly increased, the prevalence of goitre decreases from 60 to 34% as a function of a decrease in urinary SCN from 2.57 to 1.32 mg/dl.

The higher iodine intake of the populations living on or near the Zaire river probably results from their greater consumption of river fish. In the populations of Lake Tumba, which is situated near Nbandaka and fed by the Zaire river, the diet is almost exclusively based on fish. In this population, iodine supply is of the order of 100 μg/day, a much higher value than in the population of the "Gens d'eau" (unpublished observations).

Our work confirmed the important variability in the cyanide content of fresh cassava roots. The cyanide content cannot be used as a biochemical criterion for distinguishing the sweet and bitter varieties (chapter 1); nevertheless, in each village, the inhabitants were capable of recognizing the varieties that contained more or less cyanide.

The study demonstrated that nutritional habits were more closely linked to the ethnic group than to the location of the village. However, the habits were modified in semi-urban environments where *chickwangue* intake increased for groups that traditionally consumed *fuku*. This factor, added to an iodine supply that is probably slightly increased due to the diversification of the diet, explains why the prevalence of goitre was consistently lower in the urban setting than in the traditional environment, even in a zone of severe endemic goitre.

In conclusion, this work showed that the development of goitre in Ubangi is critically linked to the conditions of the nutritional microenvironment and specifically to the balance between iodine and thiocyanate. Cassava is the basic foodstuff and contains large quantities of cyanide. If it is not correctly detoxified during preparation, the food ingested contains significant amounts of residual cyanide causing an increase in thiocyanate. In the presence of iodine deficiency or even with normal amounts of iodine, thiocyanate leads to the development of goitre.

The efficiency of iodine supplements for the prevention of endemic goitre is well demonstrated (chapters 3 and 7). In this region, it is particularly important to evaluate to what extent an easily achievable improvement in the methods of preparing cassava could constitute, by itself, an effective prophylaxis against one of the most endemic diseases of the area.

Chapter 7

Mass Treatment Program with Iodized Oil

R. LAGASSE, P. COURTOIS, K. LUVIVILA, Y. YUNGA, J.B. VANDERPAS, P. BOURDOUX, A.M. ERMANS, AND C.H. THILLY

The object of this chapter is to describe the organization of a mass treatment program with low doses of iodized oil in Ubangi. The iodized oil treatment programs carried out on Idjwi and also in Equador (Ramirez et al. 1969), Peru (Pretell et al. 1969a, b, 1972), Argentina (Watanabe et al. 1974), and Greece (Malamos et al. 1970) involved limited numbers of subjects. Thus it was not known whether a program of this sort was feasible on the scale of an important public health problem nor whether it could be integrated into other programs of control of endemic diseases in developing countries. Whereas an extensive program based on iodized oil was launched among 100 000 inhabitants of Papua-New Guinea in 1972 (Hetzel 1974), the results of this program have not yet been published. In addition, large doses of iodized oil were used in the Papua-New Guinea programs, and doses were altered according to age groups and goitre types. It would not be possible, therefore, to evaluate the long-term prophylactic effect as a function of age and dose.

In our program, we planned to ensure the administration of iodized oil to the largest possible number of persons, passing in a systematic manner from village to village. Although in Idjwi no cases of secondary hyperthyroidism had been observed following iodine administration, we were particularly concerned with minimizing the risk of this complication. Recent data from Tasmania (Connolly 1973) had shown a transitory elevation of Iod-Basedow following addition of iodine to bread, and under the conditions present in Ubangi, the diagnosis and treatment of secondary hyperthyroidism would be particularly difficult. Therefore, the dose of iodized oil was reduced to 0.5 ml, i.e., approximately 200 mg iodine, and administered to all subjects regardless of age or type of goitre.

To ensure the necessary coordination of a health program involving a vast population, we created a central planning unit in Gemena responsible for the scheduling of injections and the evaluation of the results from an epidemiological, biologic, and operational point of view. In addition, we formed two mobile teams to undertake the systematic treatment of the inhabitants.

Throughout the program, particular attention was paid to the compilation of data such that a valid operational and epidemiological evaluation of the results from all villages could be performed. The following information was collected: overall prevalence of goitre, prevalence of visible and voluminous goitre, prevalence of cretinism. We calculated participation in the study by comparing the number of treated subjects with the total number of inhabitants enrolled in the census.

In September 1974, the first mobile team, consisting of one doctor and four nurses, was constituted. Table 35 shows month by month the principal epidemiological and operational determinations between October 1974 and September 1975. During this first year 86 470 people were treated, 1660–14 954 per month. This variability was principally due to the irregularity of gasoline supply and to a lesser extent to the installation of the base of operations and to the negotiations with administrative and sanitary authorities. The mean

Table 35. Principal epidemiological and operational data during the 1st year (1974–75) of the mass treatment program.

Month	Prevalence of all goitres (%)	Prevalence of visible goitres (%)	Prevalence of cretinism (%)	Number treated	Participation (est. % of population)
Oct	71	23	6.4	4798	55
Nov	73	33	2.4	2634	51
Dec	55	12	2.2	9957	60
Jan	45	6	1.6	5280	71
Feb	43	5	0.8	14954	89
Mar	35	6	0.6	12552	91
Apr	55	18	7.0	12560	84
May	60	22	10.1	9158	84
Jun	59	14	5.2	3276	78
Jul	67	20	7.0	1660	59
Aug	62	20	2.8	5245	47
Sep	70	24	9.0	4396	31

participation, 74%, was high and was probably even underestimated by 10% because of the abnormally high population estimate in the administrative census. The mean overall prevalence of goitre was 53% (35–73%), of visible goitre 14% (5–33%), and of cretinism 4% (0.6–10.1%). The lowest prevalence values were observed during January, February, and March, corresponding to the period in which inhabitants of the city of Gemena were treated.

Table 36 confirms the association between lower prevalence values and the urban environment. The most important differences between the urban and rural environments were found in prevalence values for visible goitre and cretinism, i.e., the values indicating the severity of endemic goitre. The percent participation was significantly higher in the city of Gemena where publicizing the program to encourage participation was simpler than in the rural zones. Fig. 45 shows the geographic extent of treatment in Gemena and along the principal surrounding routes during the 1st year.

In August 1975, a Zairian doctor and five Zairian nurses were recruited, and two more powerful overland vehicles were acquired. The result was a second full-time mobile team. In September 1975, a scientific colloquium was organized in Gemena for the inauguration of the central unit including a laboratory of nutritional chemistry. The national and regional authorities invited foreign experts; directors of scientific research, public health, and social services in Zaire; and Zairian, Canadian, American, and Belgian doctors, medical assistants, and public health workers involved in the subregion. During the

Table 36. Principal epidemiological data during the first 4 years of the mass treatment program, with breakdowns for urban and rural areas in the 1st year and for local and mobile health teams in the 2nd year.

Year	Prevalence of all goitres (%)	Prevalence of visible goitres (%)	Prevalence of cretinism (%)	Number screened	Participation (est. % of population)
1[a]	53 (40, 61)	14 (6, 19)	4.0 (0.8, 6.0)	86470 (32786, 54684)	74 (86, 68)
2[b]	55 (53, 59)	16 (14, 20)	4.0 (5.2, 1.8)	117895 (75849, 42046)	69 (67, 72)
3	42	14	4.6	83563	63
4	50	14	2.4	71637	72

[a]Numbers in parentheses are, respectively, urban and rural figures.
[b]Numbers in parentheses are, respectively, the numbers recorded by mobile and local health teams.

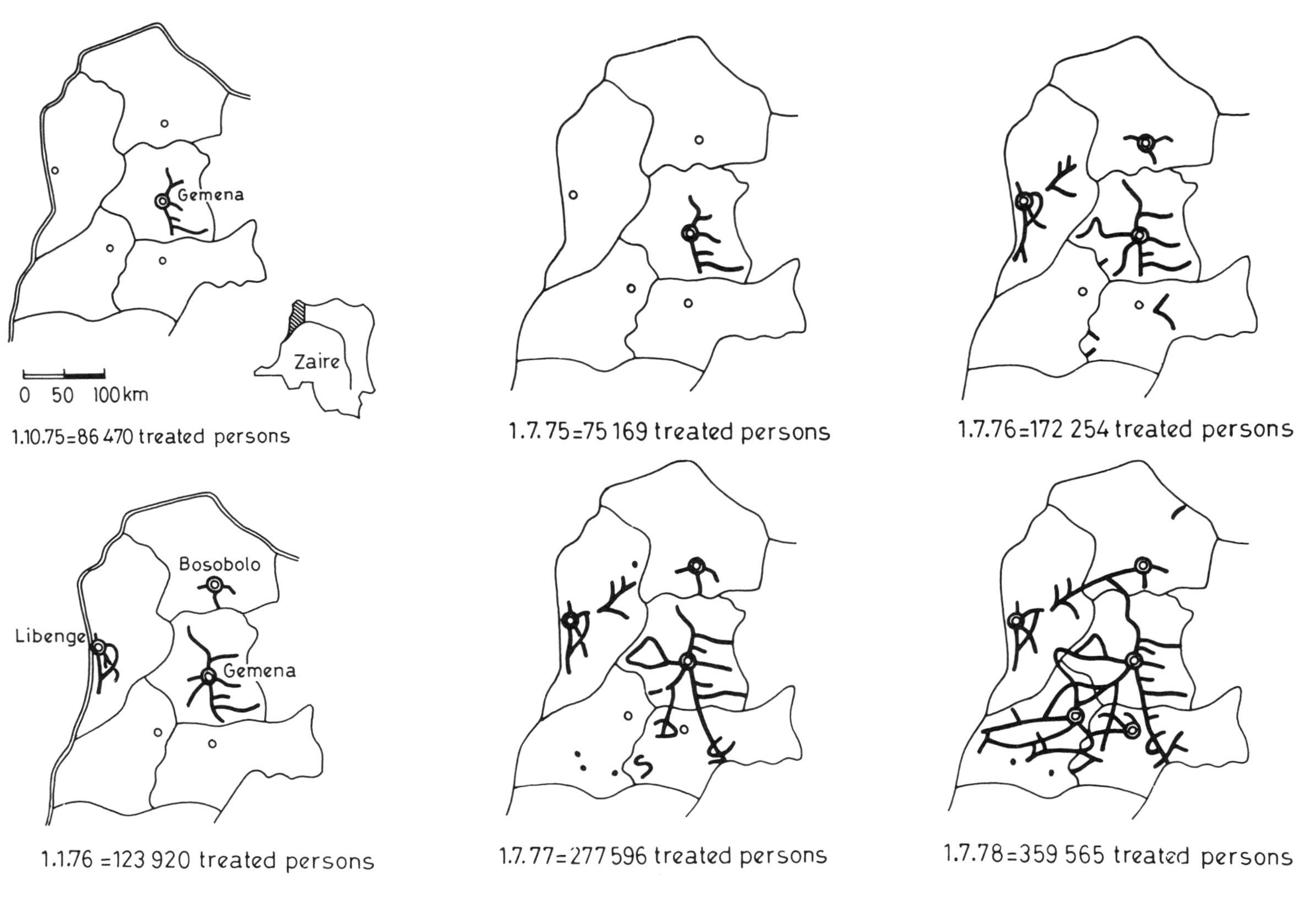

Fig. 45. *Progressive geographic extension of the mass treatment program during the first 4 years.*

colloquium, the first results of the treatment program were reported. Research on the toxicity of cassava and other health problems of the subregion and rural zones in Zaire in general was also presented. The integration of the goitre program and certain health services in rural areas of the subregion was discussed. As a result, local health teams were asked to participate in the treatment campaign in their respective zones, and two satellite operational centres were created in the small cities of Libenge and Bosobolo, thus allowing the initiation of treatment in these zones.

Table 36 also shows the principal results of the 2nd year of activity (October 1975–September 1976) during which 117895 people were treated. The results of the two mobile teams of the goitre program are separated from those of the associated local health teams. Such local teams received instruction in the central unit in Gemena. During this year the latter treated 42 046 people, and their coverage of the population was good (72%). But the epidemiological results obtained in regions of endemic goitre comparable with other rural zones differed from those of the teams integral to the program. Specifically, there was an underestimation of the prevalence of cretinism and an overestimation of visible goitre.

In another attempt to decentralize the program, the mobile units instructed rural nurses on how to give systematic treatment in the villages surrounding their dispensaries. However, the nurses were not accustomed to mass treatment programs and did not receive sufficient help from the mobile units. Thus, they did not succeed in organizing a coherent treatment program, and this attempt was abandoned.

Fig. 45 also shows the extent of the program in January 1976. The treatment zones along the principal routes surrounding Gemena, Libenge, and Bosobolo are clearly evident. During the 3rd and 4th years, the treatment program was expanded rapidly in a centrifugal manner from the three bases toward the most peripheral areas of the subregion. Fig. 45 shows this progressive geographic extension during the first 4 years. Table 36 summarizes the principal epidemiological and operational data on the 360 000 people treated during this time. In the population receiving treatment, the prevalence of goitre was 51%, of visible goitre 15%, and of cretinism 3.8%. During the 4 years, the percent participation remained high, 69% of the target population.

As the result of a meeting held early in 1977, the Zairian authorities requested that the program be consolidated and extended to other health problems, thus taking advantage of our extensive geographic coverage of the subregion. It was proposed that the mobile teams, as far as possible, carry out supervision, logistic aid, and provisioning of local medical centres and outlying dispensaries. Limited quantities of emergency medicines were dispensed to nurses in rural areas. In addition, certain of them were supervised by the mobile teams. Logistic support was provided for the hospital and hygiene laboratory in Gemena. The activities of each team were reorganized such that during each month they undertook one mission in the frame of the goitre program in the villages, the supervision of 3–10 dispensaries, and some activity in the laboratory in Gemena.

Material and Methods

Throughout the mass treatment campaign, villages were chosen at random for evaluation of the severity of endemic goitre in the area and especially for determination of the effectiveness of treatment. The effects of different doses of iodized oil administered to subjects of various ages were compared as a function of time following treatment. In addition, we compared the results in this subregion of severe endemic goitre with those obtained on Idjwi Island (Thilly et al. 1973a). The results presented here concern five villages, chosen at random within a radius of 50 km of the base in Gemena, studied between 1974 and 1977. Inhabitants of these five villages were examined before treatment and 6, 12, 18, 24, and 36 months thereafter. The subjects were divided at random into four groups receiving different doses of iodized oil, 0.2 ml, 0.5 ml, 1 ml, and 2 ml. A fifth group received an injection of vitamins, without iodine, as a placebo. The following information was gathered for all the inhabitants of the villages: the subject's and his or her parents' names, address, sex, age,

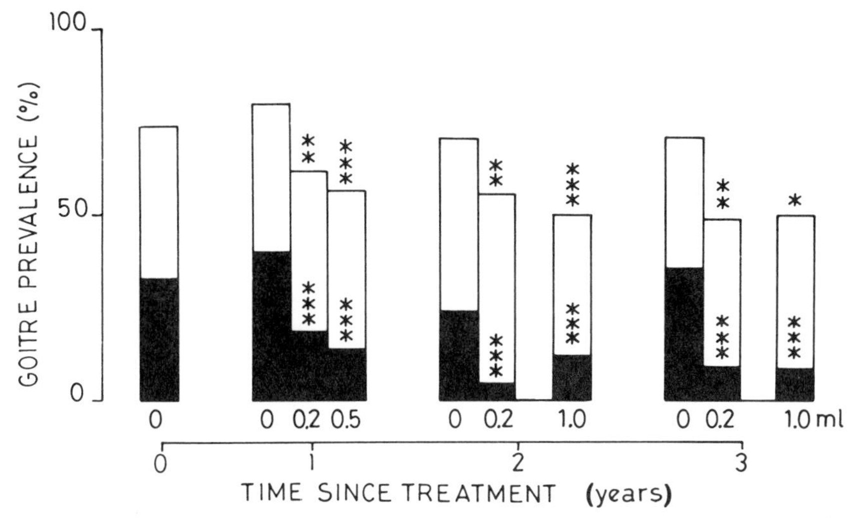

Fig. 46. *Evolution in prevalence of all goitres (white columns) and visible goitres (black columns) in untreated patients and patients treated with different doses of iodized oil (0.2, 0.5, and 1.0 ml). The results of the χ^2 test of significance comparing the placebo group with the others examined the same year are presented as: NS = not significant; * = $P < 0.05$; ** = $P < 0.01$; *** = $P < 0.005$.*

volume and nodularity of goitre, and dose of iodized oil. An identification number and individual record were assigned to each subject. From a smaller sample of subjects, we also collected blood and urine for determination of serum thyroid hormones and TSH concentrations and urinary iodide. We also measured 4-hour ^{131}I uptake and in some instances $PB^{125}I$ at equilibrium. To detect possible variations in the effectiveness of treatment as a function of age, we divided the subjects undergoing metabolic investigations into two groups — one comprising subjects younger than 25 years (youth) and the other, those older than 30 years (adults).

Table 37 shows the mean age and the number of subjects in both groups 6 months and 1 and 2 years after treatment. For practical reasons, we were unable to administer all doses or to perform all the determinations in all subjects from all villages. In particular, data for the dose of 2 ml are only available for the 6- and 18-month periods.

Results

Epidemiological data: The results of epidemiological surveys, performed in three villages before treatment in 1974 and 1, 2, and 3 years thereafter, as a function of dose are presented in Fig. 46. Before treatment, the prevalence of all goitres was 74% and of visible and voluminous goitres 33%. In the group receiving the placebo, the rates for all goitres were 80, 71, and 72% and for visible goitres 40, 24, and 36% after 1, 2, and 3 years respectively. There was some fluctuation in the prevalence rates,

Table 37. Mean age of treated subjects at different times after administration of various doses of iodized oil.

Treatment status	Youth		Adult	
	Subjects	Age (y) (mean ± SE)	Subjects	Age (y) (mean ± SE)
Untreated	57	16.3 ± 0.6	55	39.9 ± 1.1
Post-treatment				
6 months	35	14.1 ± 0.9	33	41.8 ± 2.0
2 years	22	19.2 ± 0.9	34	35.3 ± 1.0
3 years	22	14.8 ± 1.2	27	42.6 ± 1.8

but the differences were not significant, and there was no tendency to decrease. In contrast, in treated subjects, there was a clear decrease in overall prevalence, stabilizing at 50% of the population after 3 years. The differences between the placebo and treated groups were highly significant for all the doses administered. The decrease in prevalence of visible and voluminous goitres was particularly striking. The mean rate for the 3 years in the placebo group was 31%, decreasing to 11% in the subjects receiving 0.2, 0.5, or 1.0 ml iodized oil. No rebound was observed 3 years after treatment. During successive examinations, the subjects showed no clinical signs of hypo- or hyperthyroidism. There were no spontaneous or elicited complaints of side effects following treatment.

Iodine metabolism: Table 38 presents the results of measurements of thyroid function in those receiving placebo and those who had been treated 6 months, 2 and 3 years earlier with various doses of iodized oil. For the untreated subjects, the values were characteristic of endemic goitre associated with very severe iodine deficiency. On the basis of the urinary concentration, the daily iodide excretion was estimated to be 10–20 μg. The iodine deficiency was also manifested by the extreme avidity of the thyroid for iodine, as demonstrated by a mean 4-hour ^{131}I uptake of 53%. The mean serum T_4 concentration was inferior to the normal and the TSH higher than normal.

After injection of iodized oil, the subjects' urinary iodide concentration was markedly increased. After a year the values decreased in a nonlinear fashion toward the pretreatment level. At 2 years post-treatment, the daily urinary iodide excretion was approximately 50 μg/24 h. At 3 years, the urinary concentration was still 2–3 times higher than initially. The radioactive iodine uptake followed an inverse pattern with a rapid decrease followed by a slow return. The serum TSH levels fell to within normal limits (5 μU/ml) during the 3 years following treatment as did T_4 and T_3. However, the mean T_4 was markedly increased to the upper limit of normal 6 months after treatment. At 3 years post-treatment all the values of the treated group still differed significantly from those of the untreated group ($P < 0.001$).

The results of measurements of thyroid function as a function of the dose of iodized oil injected and time after treatment are presented in Fig. 47. The values in untreated subjects were clearly stable except for an isolated decrease in T_3 level at 2 years. Regardless of the dose of iodide, the mean levels of thyroid hormones and TSH were normalized and remained within the limits of euthyroidism during the 3 years of evaluation. The uptake values were drastically reduced for all doses 6 months after treatment but subsequently increased toward the initial value in all treated groups. In the group receiving the smallest dose, 0.5 ml, the uptake at 3 years was clearly elevated (41%) but still significantly different from the value in untreated subjects. The value for the other measures of thyroid function were also significantly different at 3 years with this small dose. Serum thyroxine was 5.4 μg/dl, triiodothyronine 128 ng/dl, and TSH 2.9

Table 38. Evolution of thyroid function and urinary iodide excretion as a function of time after administration of various doses of iodized oil.

Treatment status	4-h ^{131}I uptake		T_4		T_3		TSH[a]		Urinary iodide (I^-)[a]	
	Subjects	% (mean ± SE)	Subjects	μg/dl (mean ± SE)	Subjects	ng/dl (mean ± SE)	Subjects	μU/ml	Subjects	μg/dl
Untreated	85	53.0 ± 2.3	68	4.0 ± 0.4	71	188 ± 10	72	8.0 (6.0–9.5)	92	1.1 (1.0–1.1)
Post-treatment										
6 months	49	5.7 ± 0.4	49	13.0 ± 0.5	48	156 ± 5	48	1.4 (1.2–1.6)	54	58.9 (51.1–68.0)
2 years	42	21.5 ± 2.3	31	7.4 ± 0.4	48	122 ± 5	48	4.7 (4.3–5.1)	56	4.5 (3.9–5.2)
3 years	49	28.2 ± 2.3	32	7.1 ± 0.5	32	134 ± 5	31	2.4 (1.9–3.0)	48	2.5 (2.2–2.9)

[a]The value is the geometric mean ($\bar{G}$–SE,$\bar{G}$+SE).

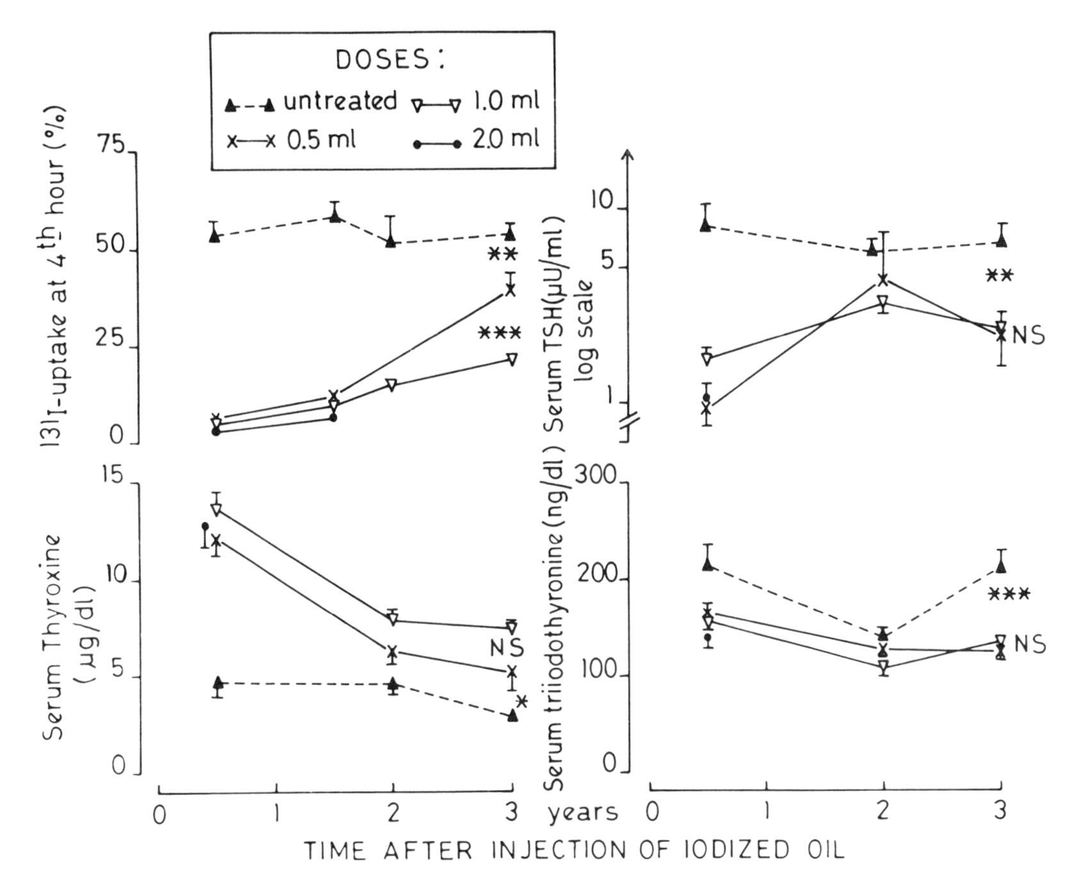

*Fig. 47. Evolution of measures of thyroid function (4-h ^{131}I uptake, T_4, T_3, and TSH) for different doses of iodized oil. The arithmetic mean ± 1 SE is shown for ^{131}I, T_4, and T_3; the geometric mean ± 1 SE for TSH. Results of Student's t-test between means are NS = not significant, * = $P < 0.05$; ** = $P < 0.01$; *** = $P < 0.005$.*

μU/ml. In comparison with this 0.5 ml group, the determinations at 6 months, 2 and 3 years following treatment with 1.0 ml were characterized by higher T_4 ($P < 0.01$, $P < 0.05$, $P > 0.05$), lower 4-hour ^{131}I uptake ($P > 0.05$, $P > 0.05$, $P < 0.001$) and T_3 and TSH that were not significantly different. When after 6 months and 18 months, the group receiving 2 ml was compared with the group receiving 0.5 ml iodized oil, the difference in values was accentuated and attained the significance level for radioactive iodine uptake and T_3. However the values for T_4 and TSH were not significantly different.

Fig. 48 presents a semilogarithmic plot of urinary iodide concentration as a function of time after treatment in the groups treated with different doses of iodized oil and in the untreated group. Low values of the order of 1–2 μg/dl were observed in the untreated group. In the group receiving 0.5 ml, the urinary iodide concentration 6 months after treatment was 20 μg/dl. The value decreased subsequently to reach 2.3 μg/dl after 3 years. In the groups receiving larger doses, the evolution was similar, the maximum value being in direct relation to dose administered. However, the differences tended to disappear with time and after 1½ years there was no significant difference between results in the groups having received 0.5 ml and those receiving 1.0 ml iodized oil.

In Fig. 49 the evolution of measures of thyroid function are presented as a function of age — subjects under age 25 years (youth) being compared with subjects over age 30

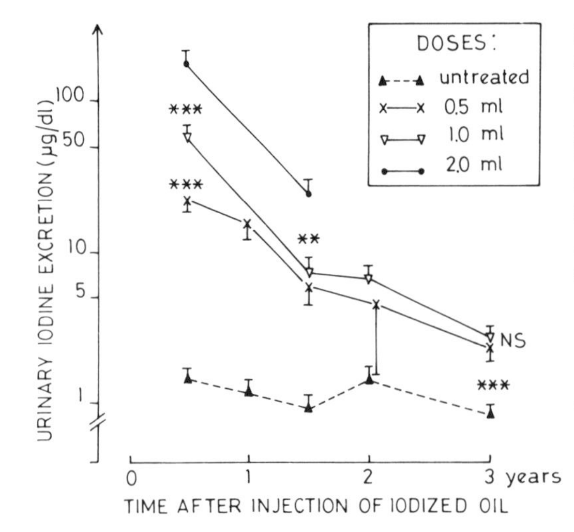

*Fig. 48. Evolution of urinary iodide concentration for different doses of iodized oil. (Geometric mean ± 1 SE; significance: NS = not significant; * = $P < 0.05$; ** = $P < 0.01$; *** = $P < 0.005$.)*

(adults). Before treatment, serum TSH and T_3 concentrations were higher in the young subjects. After treatment, ^{131}I uptake, TSH, T_4, and T_3 were all consistently higher in youth, but the differences were not significant in every instance. Six months after treatment, the mean serum T_4 in youth was very high (15 µg/dl), and the difference between the young and adult groups was highly significant.

The evolution of the $PB^{125}I$ at equilibrium and the urinary iodide concentration as a function of age are shown in Fig. 50. In young subjects, the $PB^{125}I$ was approximately double that in adults both before and after treatment. The $PB^{125}I$ was strikingly decreased 6 months after treatment and regained the pretreatment levels at 3 years. Urinary iodide values did not appear to be significantly affected by age despite a little systematic difference between the two age groups.

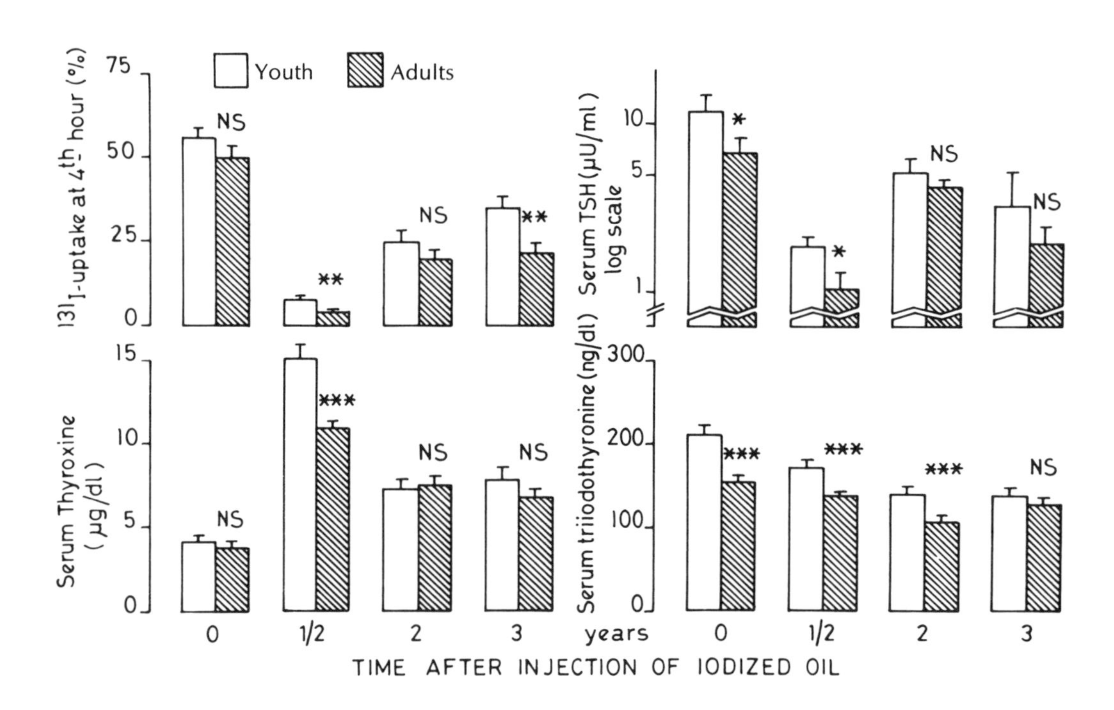

*Fig. 49. Evolution of measures of thyroid function as a function of age group. (Subjects less than 25 years old = youth and subjects more than 30 years old = adults.) The arithmetic mean ± 1 SE is shown for ^{131}I, T_4, and T_3; the geometric mean ± 1 SE for TSH. Results of Student's t-test between means are NS = not significant; * = $P < 0.05$; ** = $P < 0.01$; *** = $P < 0.005$.*

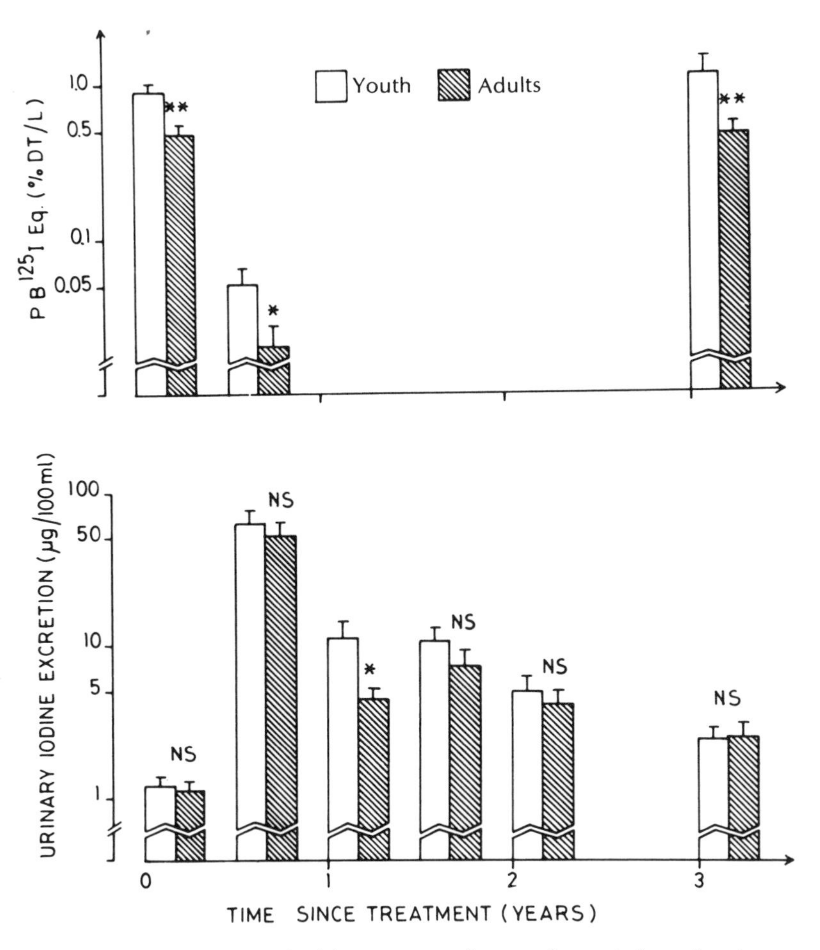

Fig. 50. *Evolution of $PB^{125}I$ at equilibrium and iodide excretion as a function of age and of time elapsed since treatment. The results are presented as the geometric mean ± 1 SE; significance: NS = not significant; * = $P < 0.05$; ** = $P < 0.01$; *** = $P < 0.005$.*

Discussion

Evaluation of the effects of iodized oil: The results presented in this chapter confirm previous data on the effect of iodized oil in New Guinea (Buttfield et al. 1967), Peru (Pretell 1972; Pretell et al. 1969a, b), Equador (Ramirez et al. 1969), and Idjwi Island (Delange et al. 1969; Thilly et al. 1973a). The present study demonstrates, in addition, that relatively small doses of iodized oil, administered to a total population, diminish the frequency of goitre, correct iodine deficiency, and normalize biochemical measures of thyroid function. These results persist over a long period. Our data indicated that in particularly severe endemic goitre, there is a clear relation between the dose administered and the subsequent changes in measures of thyroid function. Finally, there exists a relation between the biochemical indices and the age of the subjects, both before and after treatment.

In Ubangi, we observed that a low dose of iodized oil restored serum T_4, T_3, and TSH levels to normal values for at least 3 years regardless of age. Thus, the results were as favourable as those obtained with higher doses in other regions and notably on Idjwi Island. Nevertheless, a more detailed analysis of iodide metabolism revealed that, 3 years after treatment with 0.5 ml iodized oil, there was a relatively important elevation of ^{131}I uptake and particularly of $PB^{125}I$ at equilibrium, and the iodide excretion was only 2½ times as high as the value in the untreated group. In contrast, for the 1.0 ml dose, the 4-hour ^{131}I uptake after 3 years was lower, 23% of the administered dose. The serum TSH and urinary iodide were not significantly different. One can conclude that a dose of 0.5 ml is sufficient to correct iodine deficiency and biochemical indices of thyroid function over a 3-year period but that a higher dose is necessary to stabilize the situation over 5 years.

Our results also confirm previous reports that the decrease in urinary iodide concentration with time after treatment is non-exponential (Pretell 1972; Thilly et al. 1973a). In the first months, we observed a rapid decline in the amounts of iodide eliminated. Subsequently, the rate of elimination was much slower. In comparison with results from Idjwi, the urinary iodide concentrations after 3 years were five times less. This difference may have been a reflection of the larger doses injected in Kivu and/or the different environmental conditions. Daily iodine supply was twofold greater in Kivu.

The decrease in the prevalence of goitre, observed in Ubangi, was similar to that reported in Peru by Pretell and colleagues (1969a, b) but less dramatic than in Idjwi (chapter 3). This could be explained by the pretreatment epidemiological situation in Ubangi, which was more akin to that in Peru (86% prevalence of all goitres and 58% prevalence of visible goitres) than to that in Idjwi (47% all goitres and 16% visible goitres). The greater severity of endemic goitre in Ubangi (Thilly et al. 1977; chapter 4) as compared with Idjwi (Delange et al. 1969; Delange 1974) is probably related to greater exposure to goitrogenic factors (Bourdoux et al. 1978; chapters 5 and 6) and greater iodine deficiency.

Furthermore, our results demonstrate an effect of age on thyroid function. Young subjects had higher urinary iodide excretion in the months following treatment (at 1 year post-treatment, significantly different from adult values) and higher T_4, T_3, and TSH levels before and after treatment. The 4-hour ^{131}I uptake values revealed a greater avidity of the thyroid for iodine in the young group, before and after treatment. This was illustrated by the more rapid rebound of radioactive iodine uptake in this age group at 2 and 3 years. The levels of $PB^{125}I$ were systematically higher in youth. These observations confirm previous findings (Delange 1974) of accelerated functioning of the thyroid in young subjects in a region of endemic goitre and a proportional increase in iodine requirement in this age group.

The possibility of transient hyperthyroidism following iodized oil administration has already been suggested (Croxson et al. 1976; Pretell et al. 1969a, b). Young subjects, 6 months after treatment, had serum T_4 concentrations above the limits of normal; however, their serum T_3 levels were lower than in the untreated group, resisting a tendency toward thyrotoxicosis. At 2 years, the T_4 values were normalized. We did not record any complaints from the subjects, nor did we observe any abnormalities that would suggest clinical hyperthyroidism. However, a more detailed study should be undertaken to determine whether young patients should receive more frequent injections of lesser amounts. It should be noted too that in the young patients, the mean T_4 value at 6 months did not change significantly as a function of administered dose.

Organization of the mass treatment program: The results of the mass treatment program in Ubangi demonstrate the possibility of carrying out vast campaigns of goitre control. The program had limited means but progressed rapidly and was widely accepted by the target population.

The program in Ubangi was carried out by two mobile teams who administered 5000–10 000 doses of iodized oil per month over a 4-year period without interruption. The organization and execution were achieved in a particularly disadvantaged rural region of Central Africa at a time when local provisions, particularly gas, were in short supply due to world economic conditions. Thus, programs of this type can be carried out successfully in highly underdeveloped regions. Under these difficult circumstances and despite the added responsibility for other medical activities, each team performed approximately 40 000 injections per year. Under more favourable circumstances, this number could easily be increased.

The tabulation and statistical analysis of data allowed us to follow the progress of the program from village to village and to retrieve epidemiological and operational data. This documentation was essential for the preparation of the schedule of injection and reinjection of the population in such a long-term program. In New Guinea, a program of iodized oil injection involving 10 000 people was organized in 1972 (Hetzel 1974), but no provision was made for the collection of adequate statistical data to evaluate the success of the program. Thus, Ubangi is the first goitre prevention program carried out on a public health scale for which epidemiological and operational evaluation has been possible.

A major problem in developing countries is the development of primary health services. The advantage of the iodized oil treatment program is that it allows rapid treatment of goitre and protection against goitre and cretinism, which are highly visible and apparent to everyone. The spectacular nature of this action, and particularly the important reduction in visible goitres, is a strong stimulus for participation and facilitates the subsequent organization of other health programs such as vaccinations or the progressive development of stable medical facilities. Tauil et al. (1978) in a small village in the Amazon also concluded that a spectacular activity (e.g., the eradication of vermin) was necessary to achieve active participation of the community in the development of other primary care programs.

Our program made use of both stable and mobile teams. By virtue of careful and detailed instruction, local mobile teams were able to carry out the program in limited areas. In contrast, in the absence of sufficient supervision, nurses in rural dispensaries did not succeed. One must conclude that goitre prevention with iodized oil could easily be entrusted to teams involved in the attack of other endemic diseases in various African countries, but before delegating this responsibility, the health authorities must ensure further development and better organization of central units. An identical conclusion has been reached in Zaire and in other countries reviewing programs of measles prevention. Clearly, stable health units are a high priority in the Third World; however, mobile teams can provide at least a minimum of care for the most deprived populations. Because of the lack of central units, the high percentage of participation and the rapid geographic extension of our program would not have been possible without the mobile teams, nor would it have been possible if we had depended on prophylaxis with iodized salt. Not only was geographic coverage extensive but also the most remote areas received the same amount of protection as did the more favoured areas where diseases are often less severe and can be directly treated by existing local services. Thus, in a strategy to develop peripheral and rural zones, vertical programs of this type are an important means of redistributing health services from towns or more favoured rural areas to disadvantaged zones.

Our program allowed reinforcement of health services particularly at two levels: at the most peripheral level, through mobile teams, and at the intermediate level, where the central unit for planning and evaluation of such a program is normally located. Although there have been different studies on development of primary health services at the local level (Newell 1975), there have been very few published works about the necessary structures and functions at the intermediate level. In the absence of such intermediate structures, the national departments of public health often seem incapable of applying their political and budgetary decisions at the local level. Conversely, the local health services in Zaire and in Ubangi seem unable to transmit their results effectively to the central level. Taylor (1978) has recently drawn attention to the necessity of creating, at the intermediate level, logistic support and supervision for local activities.

The goitre program has resulted from a careful operational study and application of simple and specific technology whose effects are clearly measurable. The demonstrable value of such a program should encourage application of the same methods elsewhere. At present, pilot projects for the development of fundamental health services are too often characterized by a virtual absence of evaluation of the results obtained. Such an approach is likely to favour the introduction of ineffective health measures that do not benefit the population. The association, in Ubangi, of a readily evaluated program against goitre with actions whose organization and methods of evaluation are more diffuse, e.g., periodic examination of children under age 5 years, should be beneficial. Such association may foster the development of a rigorous critical sense and a better comprehension of the long-term health needs of the children and mothers of the rural areas of Ubangi among the newly created health teams.

Finally, one may ask whether a modification of dietary habits, specifically a decrease in cassava consumption, would lead to a more lasting control of endemic goitre and cretinism in Ubangi. The answer to this question is unknown and merits further investigation. It is clear that readministration of iodized oil at regular intervals does not represent a definitive solution to the problem of endemic goitre but rather a first attack, which has been spectacular and effective. Other methods of eradication must now be found, ensuring a long-term solution of the problem.

Chapter 8

Studies of the Antithyroid Effects of Cassava and of Thiocyanate in Rats

A.M. ERMANS, J. KINTHAERT, M. VAN DER VELDEN, AND P. BOURDOUX

In this chapter, we have reported a series of experimental investigations carried out in rats between 1972 and 1978. The first part is devoted to studies on rats fed a low iodine diet (LID), comparing the long-term effects of consuming cassava with the effects of receiving graded doses of thiocyanate. The second part deals with the short-term effect induced by the ingestion of a single cassava meal on the serum thiocyanate concentration and its consequences on the thyroid uptake of radioactive iodide. The third part deals with the modifications of the iodide pump induced by a moderate rise of the SCN serum level in rats chronically supplemented with low doses of thiocyanate. The fourth part tests the hypothesis that an increased manganese supply could induce an antigoitrogenic effect.

Antithyroid Effects of Cassava and Thiocyanate

Cassava is one of the basic foodstuffs in many tropical countries (Nestel 1973). In 1966, Ekpechi et al. reported that administration of crude cassava tubers to iodine-deficient rats provoked a marked hyperplasia of the thyroid gland and suggested that this effect was related to a component of the thionamide group. Delange and Ermans (1971) observed that ingestion of a large meal of cassava flour by inhabitants of a goitrous area in Central Africa induced a significant drop of the ^{131}I thyroid uptake and a concomitant rise of the urinary excretion of ^{131}I and ^{127}I; abnormally high concentrations of thiocyanate were found in plasma and in urine of this population (Delange 1974). It was therefore suggested that the goitrogenic action of cassava could be caused by a thiocyanate-like substance (Delange and Ermans 1971). Moreover it was reported that cassava ingestion in rats markedly increased SCN concentration in plasma and induced a severe depletion of the thyroid iodine content (Ermans et al. 1972, 1973).

Nevertheless no detectable amounts of thiocyanate nor of its precursors have been detected so far in cassava (Montgomery 1969). Cassava contains cyanogenic glucosides, mainly linamarin, from which autohydrolysis liberates large amounts of cyanide; conversion into thiocyanate is known as the principal metabolic pathway ensuring endogenous cyanide detoxification (van Etten 1969).

The question therefore arises whether endogenous production of thiocyanate secondary to the absorption of cyanogenic glucosides from cassava accounts for the goitrogenic effect observed after the ingestion of cassava. The purpose of the present study was to answer this question in rats; to this end we have evaluated the modifications of iodine and SCN metabolism induced by chronic administration of cassava and of graded doses of thiocyanate.

Antithyroid effects of SCN are well documented (Funderburk 1966; Halmi 1961; Maloof and Soodak 1964; Scranton et al. 1961; Wolff 1964; Wollman 1962); SCN ion inhibits the iodide pump and competes with iodide in the organification processes; however all studies carried out so far mainly concern the acute effects of SCN^-. In a preliminary study, we reported an unexpected, small increase of the plasma SCN^- levels after prolonged overload with huge amounts of SCN^- (Van der Velden et al. 1973). We paid particular attention therefore to the comparison of the effects induced by acute and chronic administration of thiocyanate.

Materials and methods

Five successive investigations (A to E) were carried out, each one on 75–120 rats. Acute effects of SCN were studied in experiment A, chronic effects in experiments B, C, D, and E, and the influence of cassava in experiments B, C, and D. The parameters investigated were:

- SCN^- distribution: SCN^- plasma concentration, SCN^- urinary excretion. $[^{35}S]$ SCN^- disposal rate and SCN^- renal clearance; and
- Iodine metabolism: ^{127}I content, ^{131}I uptake, and ^{125}I-labeled iodoaminoacids distribution in the thyroid gland, plasma $PB^{127}I$, and iodide renal clearance. Some of these parameters were studied in rats fed on diets supplemented with iodine.

A total 383 male Wistar rats weighing 100–160 g were used. The specific diets given to each subgroup (9–25 rats) were administered without change for 2–5 weeks until the animals were sacrificed, at which time, they were exsanguinated by cardiac puncture under diethylether anesthesia. The thyroid glands were excised, cleaned superficially, weighed, and frozen. The animals had been fed an iodine-poor diet (Remington). SCN supplements (0.1–10 mg/day) were added to the diet before cooking. Cassava was given in the form of small pieces of crude tubers, which were added to the Remington diet. Crude cassava tubers were imported from the endemic goitre area of Idjwi Island in Zaire (Delange and Ermans 1971), stored at -20 °C until they were used; tubers were peeled immediately before administration; after hydrolysis their cyanide content was about 160 mg/kg (Van der Velden et al. 1973). Iodine supplement (10 μg/day) was added to the drinking water.

The chronic administration of SCN^- or of cassava does not significantly modify the weight of the animals as compared with controls.

Meals were given once a day in the morning; the investigations and the killing of the animals were carried out in the morning or at the beginning of the afternoon. Because the food was ingested mostly during the night, determinations were generally done during the postabsorptive phase.

We carried out experiment D with the aim of estimating renal and thyroid clearances for definite plasma levels of SCN^-. In this particular investigation, animals were accustomed to being fed only between 900h and 1600h, and therefore we performed our final investigations during the same period to obtain determinations during SCN^- absorption.

^{131}I thyroid uptake: We measured ^{131}I thyroid uptake 4 hours after injecting the animals with 10 μCi carrier-free ^{131}I (CEN, Belgium) intraperitoneally. We used the gamma well-type scintillation counter (Philips, Netherlands).

^{125}I-labeled iodoaminoacids content of the thyroid gland: Rats were injected intraperitoneally with a single dose of 10 μCi carrier-free ^{125}I (CEN, Belgium) 5 weeks before they were killed, immediately before the administration of the specific diets. Homogenization of thyroid tissue, extraction of the soluble fraction, and hydrolysis by pronase were carried out as described elsewhere (Ermans et al. 1968). Separation of iodoaminoacid was achieved by thin-layer chromatography (Delange et al. 1972) on cellulose plates of 0.1 mm thickness (DC Alufolien Cellulose Merck); the solvents used were the organic phase of the following mixtures: N-butanol•acetic acid•H_2O (16:1:3) for the separation of monoiodotyrosine (MIT) and diiodotyrosine (DIT) and methanol •hexane •tertiary amyl alcohol •NH_4OH 2N (1:1:5:6) for the separation of T_4 and T_3 (Delange et al. 1972b).

^{127}I content of the thyroid gland and plasma $PB^{127}I$: Measurements of stable iodine were done

following the methods of Barker et al. (1951) using a Technicon Autoanalyzer (Chauncey, New York).

SCN measurements: To measure SCN concentrations, we used Aldridge's method (1945) for the urine samples and the method of Michajlowskij and Langer (1958) for the plasma samples.

SCN disposal rate: We injected rats with 20 μCi [35S]SCN⁻, specific activity of 130 μCi/mM (CEN-Mol, Belgium) intraperitoneally, then killed individual animals after 6, 12, 24, 30, and 48 hours; according to the faster turnover-rate observed in the presence of large SCN⁻ supplements, a shorter schedule (2, 4, 6, 8, and 24 h) was applied to rats receiving a SCN⁻ dose higher than 0.2 mg/d.

^{35}S activity in serum and urines was measured in a liquid scintillation counter (Nuclear-Chicago Mark I) after digestion of 50 μl serum or urine in 1 ml Soluene (Packard) and addition of 15 ml toluene containing omnifluor (NEN Chemicals).

The possible conversion of [35S]SCN⁻ to [35S]SO_4 (Maloof and Soodak 1959) was tested by thin-layer chromatography of serum. The investigation included 17 samples of serum obtained 6–48 hours after injection of the tracer. The separation of SCN⁻ and SO_4 was accomplished by ascending thin-layer chromatography on a cellulose plate of 0.1 mm thickness (D.C. Alufolien Cellulose Merck). The solvent was ethanol•pyridine•H_2O•30% (w/v) NH_4OH (30:30:40:2.5).

Measure of iodide and SCN clearances: The thyroidal and the renal clearances of iodide and the renal clearance of SCN⁻ were measured on three groups of 25 animals receiving respectively a supplement of 0, 1, and 5 mg SCN⁻ per day for 2 weeks. In each group, four rats were killed respectively 15, 60, 150, and 240 min after ^{131}I injection (t = 0) and 9 rats after 330 min.

Plasma concentrations of [^{131}I] iodide (% of dose/ml) and SCN⁻ (mg/dl) were measured, and the mean values were plotted against time on a diagram. The curves representing the evolution of these concentrations for each investigation were drawn; the integral was calculated from 0–330 min and the values obtained were divided by 330.

The value thus obtained represents the plasma level of [^{131}I]iodide (C_I) or of thiocyanate (C_{SCN}), if a stable concentration is assumed for the whole 0–330 min. Rats killed after 330 min were kept in individual cages between the time of ^{131}I injection and sacrifice, and urine was carefully collected. Amounts of SCN⁻ (mg) and ^{131}I (% of dose) recovered in these urine samples respectively corresponded to Q_1 and Q_2; ^{131}I percentage taken up by the thyroid gland after 330 min was Q_3. Clearances were calculated from the following equations:

Q_1/C_{SCN} = the renal clearance of SCN⁻
Q_2/C_I = the renal clearance of iodide
Q_3/C_I = the thyroid clearance of iodide

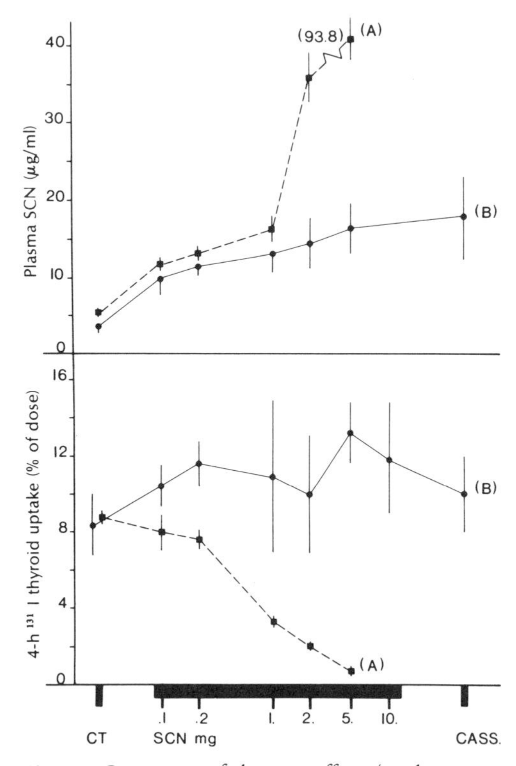

Fig. 51. *Comparison of the acute effects (single injection: dotted line) and of the chronic effects (chronic ingestion: solid line) of graded amounts of thiocyanate (0.1 to 10 mg) and of crude cassava tubers (5 g/d) on plasma SCN concentration and on the 4-h ^{131}I thyroid uptake. All rats were fed a Remington diet supplemented with 10 μg I/d. Each value is the mean observed in five rats. Vertical lines correspond to the standard deviation.*

Results

SCN^- plasma concentration and ^{131}I thyroid uptake: In Fig. 51 the SCN^- plasma concentrations and ^{131}I thyroid uptake observed after single doses of SCN (experiment A) are compared with those observed when similar doses are administered by mouth daily for 3 weeks (experiment B). In the two experiments, similar, moderate increments in SCN^- concentration were observed at low doses (0.1 and 0.2 mg); however at doses ranging from 1 to 10 mg the results of the two experiments diverged markedly, the chronic-dose group continuing to manifest slight increments and the single-dose group exhibiting a sharp rise.

The rats receiving single injections of SCN^- had decreased ^{131}I thyroid uptake in direct relation to the administered doses, whereas the chronically supplemented rats showed no inhibition of ^{131}I uptake over the entire range of SCN^- doses. In fact, despite the important scatter of experimental data, a clear-cut increase of ^{131}I uptake was apparent; differences were statistically significant for the 0.2, 1.0, and 5.0 mg doses ($P < 0.001$).

The finding that chronic administrations of thiocyanate — even in huge amounts — failed to induce a marked increase of SCN^- plasma levels was also evidenced in experiments C and D (Fig. 52).

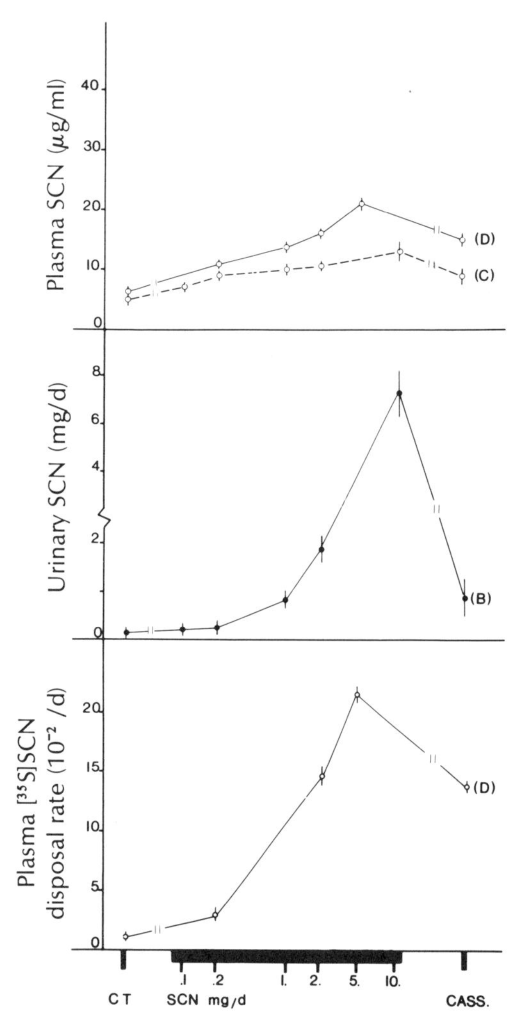

Fig. 52. Modifications of various parameters of the thiocyanate metabolism induced by the chronic administration either of graded doses of SCN or of cassava.

Distribution and kinetics of SCN^-: In control rats, SCN^- daily urinary excretion only reached 0.06 ± 0.01 mg; however, the value was considerably higher in rats chronically supplemented with graded doses of SCN^- (Fig. 52). Recovery of the administered doses that were equal or greater than 1 mg ranged from 75 to 95% (mean 83%).

When $[^{35}S]SCN^-$ was administered to control rats and rats chronically treated with 0.2 and 2 mg SCN^- per day, the time it took to disappear from their plasma was linear when plotted on semi-log coordinates for all doses of SCN^- tested (Fig. 53). This finding was used to compute the disposal rate (Fig. 52). In control rats, this rate reached 0.014/h; it progressively increased with increasing doses of SCN^- and reached values 20–40 times as high in the largest doses (Fig. 52).

The chromatographic analysis of serum samples used for these measurements showed that 98–100% of the ^{35}S measured represented SCN^-. This finding validated the use of the $[^{35}S]SCN^-$ disposal rate to compute the plasma SCN^- turnover.

Renal and thyroid clearances: In experiment D, the diet containing the thiocyanate doses of 1 and 5 mg/day was entirely administered during a restricted period of time (about 7 hours). Table 39 shows the mean plasma concentrations of SCN^- and $[^{131}I]$iodide during this time (330

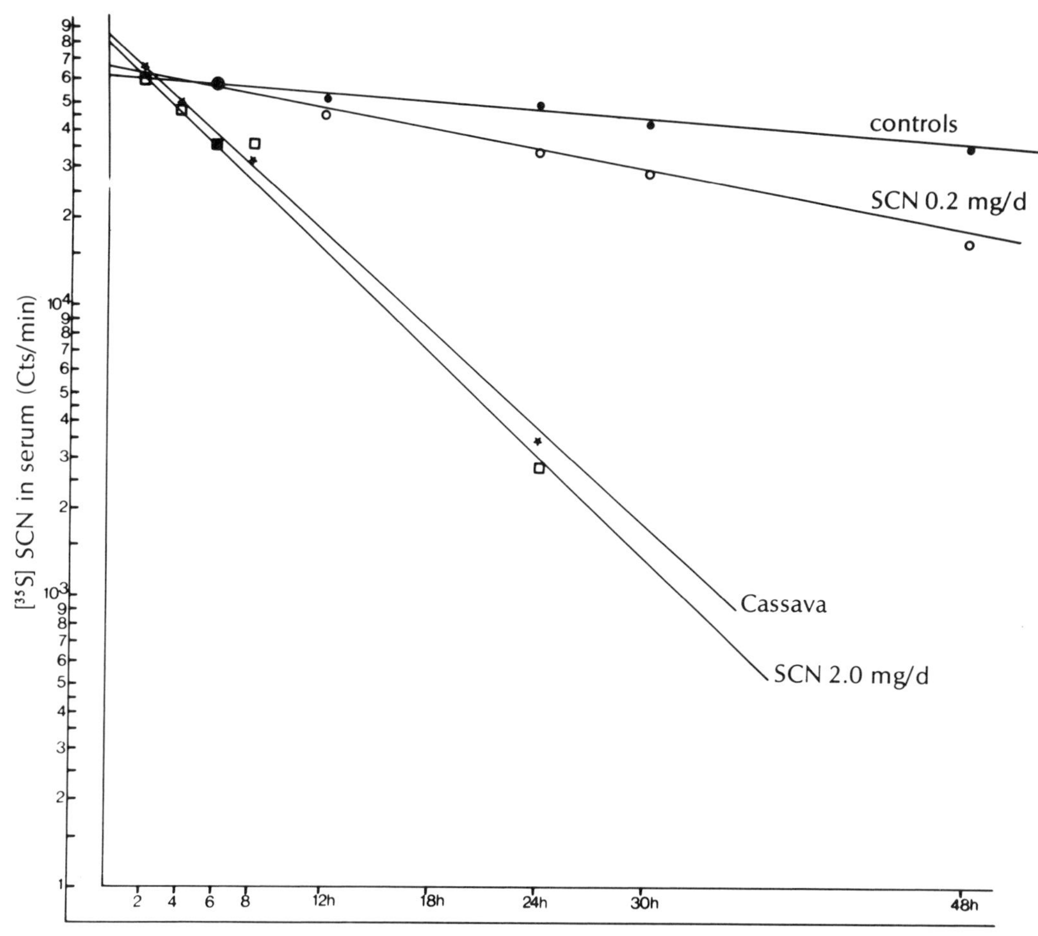

Fig. 53. *Disposal rates of* [35S] *SCN injected in control rats and in animals chronically supplemented either with thiocyanate (0.2 and 2 mg/d) or with cassava.*

Table 39. Measurements of thyroid and renal clearances of thiocyanate and iodide after prolonged supplementation with thiocyanate.

Parameter	Control rats[a]	Rats supplemented with SCN^- 1 mg/d	Rats supplemented with SCN^- 5 mg/d
SCN^- plasma concentration (μg/ml) (mean ± SD)	5.3 ± 0.6[b]	11.4 ± 2.0**	23.4 ± 4.1**
Computed mean, SCN^- concentrations (C_{SCN}) (μg/ml)	5.2	11.6	22.8
SCN^- renal clearance (10^{-2} ml/min) (mean ± SD)	0.4 ± 0.1	4.2 ± 1.2**	15.0 ± 2.9**
Iodide ^{131}I plasma concentration (% of dose/ml) (mean ± SD)	1.29 ± 0.56	1.24 ± 0.50	1.36 ± 0.61
Computed mean, [^{131}I] iodide concentrations (C_I) (μg/ml)	1.38	1.35	1.55
Iodide renal clearance (10^{-2} ml/min) (mean ± SD)	6.7 ± 0.8	8.2 ± 1.5	9.3 ± 0.1*
4-h ^{131}I thyroid uptake (% of dose) (mean ± SD)	5.4 ± 1.2	5.6 ± 0.7	4.3 ± 1.0
Iodide thyroid clearance (10^{-2} ml/min) (mean ± SD)	1.5 ± 0.2	1.4 ± 0.3	1.2 ± 0.2

[a]Significance: * = $P < 0.01$; ** = $P < 0.005$; *** = $P < 0.001$.
[b]Standard deviation of concentration levels observed at 0–330 min.

min). The magnitude and the standard deviation of the plasma SCN^- concentrations showed adequate stability at each level of SCN^- supplementation during the determinations of renal and thyroid clearances.

In control rats, the mean renal clearances of SCN^- and iodide were respectively 0.4 and 6.7 $\times 10^{-2}$ ml/min. In rats receiving 1 and 5 mg SCN^- per day, renal SCN^- clearances increased respectively by a factor of 10 and 40. Renal clearances of iodide were moderately increased and statistically significant only for the dose of 5 mg ($P < 0.01$). For this dose the thyroid clearance and the thyroid ^{131}I uptake were moderately reduced.

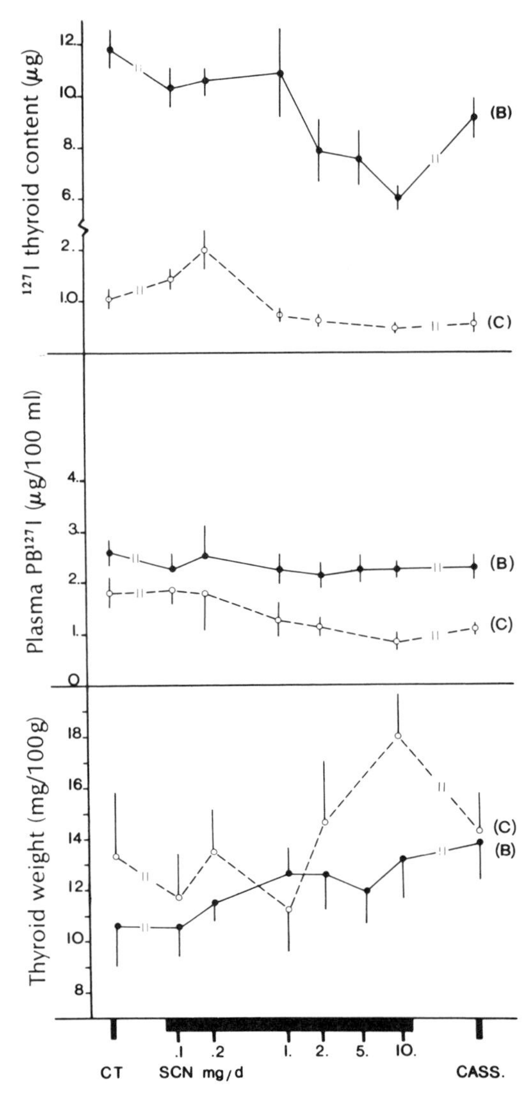

Fig. 54. Modifications of various parameters of thyroid function induced by chronic administration of graded doses of SCN and of cassava. In experiment B, the Remington diet was supplemented with 10 µg I/d.

Modifications of iodine metabolism: Chronic administration of 1–2 mg/day of SCN^- provoked a marked reduction of the thyroid iodine content in rats receiving diets both iodine-supplemented and not (Fig. 54). The degree of depletion became more severe with increasing SCN^- doses and, at the 10 mg dose, reached 40–50% of the control value. It is noteworthy that, in iodine-deficient rats, the iodine content was paradoxically increased by administration of low doses of SCN^-.

Plasma $PB^{127}I$ was not significantly reduced in rats supplemented with iodine whatever the doses of SCN^- administered (Fig. 54); however, in iodine-deficient animals it was significantly decreased at large doses of SCN^-. In comparison with the control group, SCN^- overload increased thyroid weight in iodine-supplemented rats (Fig. 54), but the increase was only significant for the largest doses ($P < 0.01$).

Fig. 55 shows the influence of increasing doses of SCN^- on the long-term distribution of ^{125}I among thyroidal iodoaminoacids in iodine-deficient rats chronically fed with graded doses of SCN^-. In control animals and after administration of low doses of SCN^- (0.1 and 0.2 mg/d), the MIT/DIT ratio ranged to about 1:1; T_3 and T_4 fractions respectively represented 10 and 25% of the whole thyroidal ^{125}I. For larger doses of SCN^-, the MIT/DIT ratio increased to about 2:1 for the 10 mg dose; in contrast the thyroxine content strongly decreased; the T_3 fraction was not significantly altered. In iodine-supplemented rats the MIT/DIT ratio was reduced as was the content of the labeled T_3 in the gland; the T_4 content was unchanged (Fig. 55).

Antithyroid action of cassava: The results referring to the action of cassava are arranged here in a distinct section; they were obtained in the same investigations as those concerning the action of SCN^-.

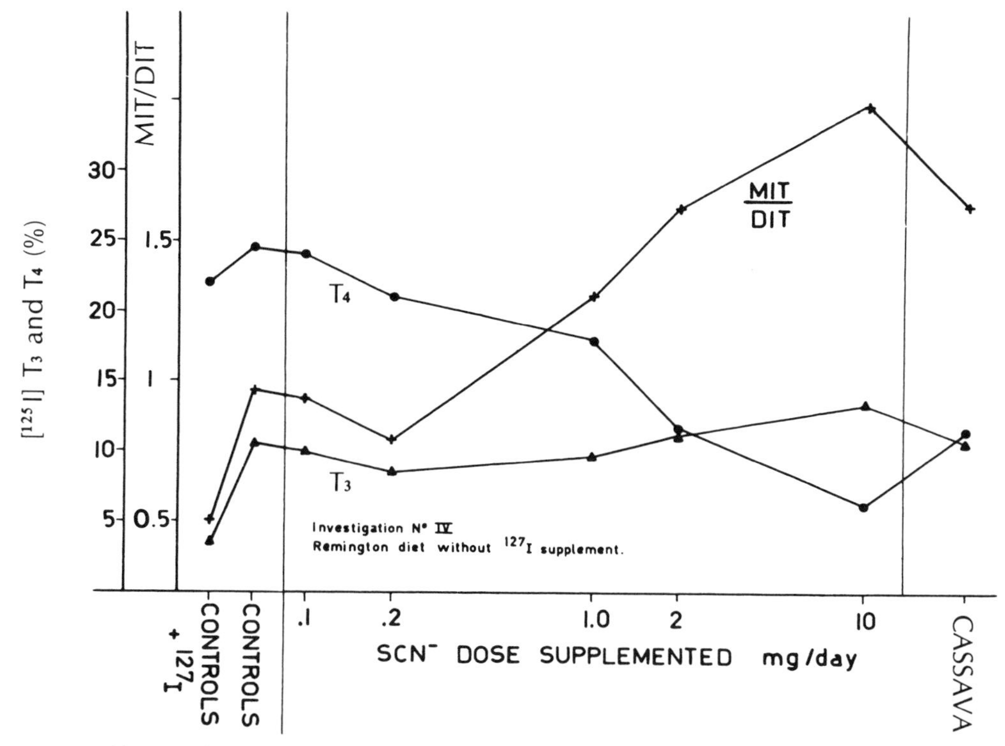

Fig. 55. *Modifications of the [125I] MIT/[125I] DIT ratio and of the [125I] T_4 and T_3 content of thyroid glands in rats given a low-iodine diet supplemented either with 127I (10 μg/d) with graded doses of thiocyanate or with cassava (manioc).*

In three distinct investigations (B, C, and D), prolonged ingestion of 10 g of crude cassava tubers per day provoked a significant but moderate increase of SCN^- plasma concentrations (Fig. 51 and 52). Fig. 52 and 53 show that cassava induced an acceleration of the disposal rate of plasma [35S]SCN^-, which was enhanced by a factor of 16 as compared with the control value. Moreover, cassava ingestion induced an increment of the urinary excretion of SCN^-, which reached 0.84 mg/d compared with 0.06 mg/d in the control rats. Modifications of SCN^- distribution related to cassava intake (10 g/d) quantitatively corresponded (Fig. 52 and 53) to those induced by the administration of doses of SCN^- amounting to about 1–2 mg/d.

Prolonged intake of cassava also produced a distinct thyroid hyperplasia in rats supplemented with iodine (Fig. 54). No inhibitory effect of thyroid 131I uptake was observed (Fig. 52).

A marked reduction of thyroid iodine content occurred in rats fed a diet with/without iodine supplementation (Fig. 54). In iodine-deficient rats, cassava ingestion induced clear-cut modifications of the 125I distribution among the iodoaminoacids of the thyroid gland.

The modifications were qualitatively similar to those observed after chronic administration of thiocyanate, i.e., increased MIT/DIT ratio, decreased T_4 fraction, and unchanged T_3 fraction. In the same animals, a significant decrease of plasma PB127I was also observed (Fig. 54).

From a quantitative point of view, all modifications of iodine metabolism induced by cassava could be superimposed on the ones obtained in the same investigations in rats supplemented with SCN^- for doses of about 1–2 mg/d (Fig. 54 and 55).

Discussion

Renal adaptation mechanism during chronic SCN^- overload: Prolonged administration of SCN^- in rats failed to increase markedly the plasma SCN^- levels as expected on the basis of the findings obtained after single injections of SCN^- (Delange and Ermans 1971). Our present observations indicated that in chronic conditions large SCN^- supplements are very rapidly excreted in urine because of drastically increased renal clearance of SCN^-. The efficiency of this adaptation is such that only slight increases in plasma SCN concentration occur no matter how large the SCN overload. In plasma, only the acceleration of $[^{35}S]SCN^-$ disposal rate allows one to give a correct estimation of the size of the SCN^- supply.

The renal adaptation mechanism probably reflects the inability of the kidney tubules to reabsorb large amounts of filtered SCN^-; this possibility is in agreement with observations of Funderburk and van Middlesworth (1971) who showed that in rats large doses of SCN^- have a faster rate of disappearance from the plasma than does endogenous SCN^- at normal concentrations.

Alterations of iodine metabolism after chronic SCN^- overload: The lack of marked increase of SCN^- concentration in plasma explains the absence of inhibition of the thyroid pump observed after SCN^- supplementation. Increased ^{131}I uptake found in this condition probably reflected a compensatory stimulation by TSH in response to the severe depletion of the iodine stores; this depletion also took place in iodine-supplemented rats (Fig. 54).

Abnormalities of the labeled iodoaminoacids induced by prolonged SCN^- overload, i.e., increased MIT/DIT ratio, decreased T_4, and unaltered T_3 fractions, are typically those found in severe iodine deficiency (Barnaby et al. 1965; Studer and Greer 1968; Inoue and Taurog 1968). Such modifications of intrathyroidal iodine metabolism have been shown to be related to abnormally low levels of iodination of thyroglobulin (Inoue and Taurog 1968). In agreement with this view it may be seen that the progressive degree of severity of the abnormalities induced by increasing doses of SCN^- (Fig. 55) strictly coincided with a corresponding decrease of the iodine content of the glands (Fig. 54).

The question arises by which mechanism prolonged SCN^- administration induces the depletion of the iodine content of the gland in the absence of any evidence of an inhibitory effect on the iodide pump. Our observations are in agreement with the view that iodine depletion could be partly caused by the enhancement of the renal clearance of iodide. This finding could be explained by a competitive action of SCN^- ion on the tubular resorption of iodide. Data suggesting such a competition of both ions at the renal level have been reported by Halmi et al. (1958) in the rat.

Modifications of the SCN^- and iodine metabolisms induced by cassava: Prolonged absorption of cassava produced in rats modifications of the SCN^- kinetics. Taking into account the amounts of SCN excreted in urine and the value of the plasma $[^{35}S]SCN^-$ disposal rate, we deduced that the daily absorption of 10 g cassava induced the endogenous production of a quantity of SCN^- ranging to 1–2 mg/day, i.e., 10–20 times as much as the daily production of SCN^- observed in the control animals.

As mentioned earlier (Van der Velden et al. 1973), neither SCN^- nor precursors of SCN^- can be detected in cassava, but endogenous production of this ion may result from the conversion of the large amounts of cyanide found in cassava. Van der Velden et al. (1973) computed from the daily absorption of 10 g cassava (in which cyanide concentration was 160 mg/kg) that complete conversion of cyanide to SCN^- would produce daily 3 mg SCN^-. This mechanism could thus quantitatively account for the production of SCN^-, i.e., 1–2 mg/day.

Our observations confirmed the goitrogenic action of cassava in rats. All effects of cassava intake on iodine metabolism were identical to those found after the administration of SCN^- at qualitative and quantitative points of view: iodine depletion, modifications of the iodoaminoacids distribution of the thyroid gland, decreased plasma $PB^{127}I$, and lack of inhibition of the iodide pump. The degree of the various anomalies also fitted with a daily thiocyanate production of about 1–2 mg/day.

In conclusion, the anomalies simultaneously induced on the metabolism of iodine and of SCN^- after chronic ingestion of cassava were so similar to those produced by well-defined amounts of thiocyanate that we accept the hypothesis that the antithyroid action of cassava is due to the endogenous production of SCN^-. In these experimental conditions, the essential action of cassava, as that of SCN^-, was to induce an iodine depletion that can be partially corrected by iodine supplementation. It is noteworthy that this antithyroid action can only be distinguished from the effects of iodine deficiency alone by assessing specific parameters of SCN^- metabolism. In the absence of such information, the goitrogenic action of some diets, which in the past has been attributed to their low iodine content, may be caused by thiocyanate-like substances largely distributed in vegetal foodstuffs (van Etten 1969).

Short-Term Action of Cassava Ingestion on SCN^- Plasma Levels in Rats

Observations made by Delange and Ermans (1971) showed that ingestion of a meal flour made from cassava grown in Idjwi Island brought about a distinct drop in thyroid uptake of radioactive iodine and an increase in its renal excretion in humans. Ekpechi (1967) and Van der Velden et al. (1973) showed that the antithyroid action of cassava observed in rats chronically fed this food was related to the endogenous production of thiocyanate from cyanide released from cassava. Chronic ingestion of cassava in rats increases the plasma thiocyanate concentration and the thiocyanate excretion and depletes the iodine content of the thyroid gland. A daily administration of 1–2 mg thiocyanate produces the same effects (Van der Velden et al. 1973).

We undertook this study to expand earlier findings; our objectives were:

- First, to mimic in rats the short-term antithyroid action of cassava observed in humans and to define the relation between the inhibition of the thyroid pump and the increase of the plasma thiocyanate concentration.
- Second, to study the temporal relation between the process of cyanide detoxification and the increase of the plasma thiocyanate concentration.

Materials and methods

To reach our goals we carried out two experiments. In the first, we fed 84 male Wistar rats weighing 170 g (mean weight) a low-iodine test diet (Nutritional Biochemicals Corporation) for 4 weeks. Then we grouped them into 21 rats each. In each group, 18 rats were force-fed. Group 1 received 2 ml distilled water; group 2, 2 ml thiocyanate (1 mg); group 3, 2 ml thiocyanate (2 mg); and group 4, 5 g cassava roots ingested in ½ hour (after 48 hours fasting).

Immediately after force-feeding, all rats were injected intraperitoneally with [^{35}S] thiocyanate (45 μCi) and ^{131}I (3 μCi) (CEA). In each group, three rats were used as controls and were neither force-fed nor injected. The control rats of group 4 fasted 48 hours but were neither force-fed nor injected. In each group, rats were killed 0.5, 1, 2, 4, 6, and 8 hours after injection (three rats at each time and in each group). Rats were exsanguinated under diethylether anesthesia. Blood was collected, and the thyroid glands were excised.

Rats killed 8 hours after injection were placed in metabolism cages immediately after injection, and their urine was collected. Thiocyanate in serum and urine was estimated according to the method of Aldridge (1945) modified by Michajlowskij and Langer (1959). Iodine-131 uptake by thyroid glands was measured in a gamma-well type scintillation counter (Philips, Netherlands). Sulfur-35 activity in serum and urine was measured in a liquid scintillation counter (Nuclear-Chicago Mark I); 100 μl serum was digested in 1 ml Soluene (Packard) and then 15 ml scintillating toluene containing 4 g omnifluor per litre (NEN Chemicals) was added.

In our second experiment, 27 male Wistar rats weighing 190 g (mean weight) were injected intraperitoneally with 250 μCi [^{14}C]cyanide (CEA). They were successively killed 2, 5, 10, 15, 30, 60, 120 minutes, and 24 hours after injection (3 rats at each time) by exsanguination under diethylether anesthesia. Blood was collected and heparinized. Thyroid gland, brain, and liver were excised and weighed.

Heparinized blood was centrifuged, and plasma separated from erythrocytes. Plasma and erythrocytes were then precipitated with an equal volume of 15% (w/v) trichloroacetic acid.

We used supernatants to perform ascending thin-layer chromatography, separating [^{14}C]-cyanide from [^{14}C]thiocyanate on aluminum cellulose sheets (Merck) of 0.1 mm thickness in the solvent, ethanol • pyridine • H_2O • 30% (w/v) NH_4OH (30:30:40:2.5). Thin-layer chromatograms were then scanned for ^{14}C activity with a Berthold Dünnschicht Scanner II at a speed of 120 mm/h. We localized the [^{14}C]-cyanide and [^{14}C]thiocyanate by comparing them with a chromatogram of a mixture of [^{14}C]cyanide and [^{35}S]thiocyanate; 100 μl of each supernatant was digested in 1 ml Soluene (Packard) before the addition of 15 ml scintillating toluene containing 4 g omnifluor (NEN Chemicals) per litre. We measured ^{14}C activity in a liquid scintillation counter (Nuclear-Chicago Mark I). Tissues were homogenized in 500 μl 15% (w/v) trichloroacetic acid (two lobes for the thyroid, 100 mg for brain and liver tissues) then centrifuged. Using a liquid scintillation counter (Nuclear-Chicago Mark I), we measured ^{14}C activity in the supernatants (100 μl supernatant digested in 1 ml Soluene (Packard) and combined with 15 ml scintillating toluene).

Results

Plasma thiocyanate concentration and ^{131}I-uptake inhibition: Oral administration of a single dose of thiocyanate rapidly increased the plasma thiocyanate concentrations in the rats to high levels: 48.9 ± 1.3 and 23.0 ± 2.6 μg/ml respectively for 1 and 2 mg thiocyanate; the plasma concentrations then rapidly decreased but were still higher than 10 μg/ml after 8 hours. Ingestion of 5 g cassava tubers increased the plasma thiocyanate concentration progres-

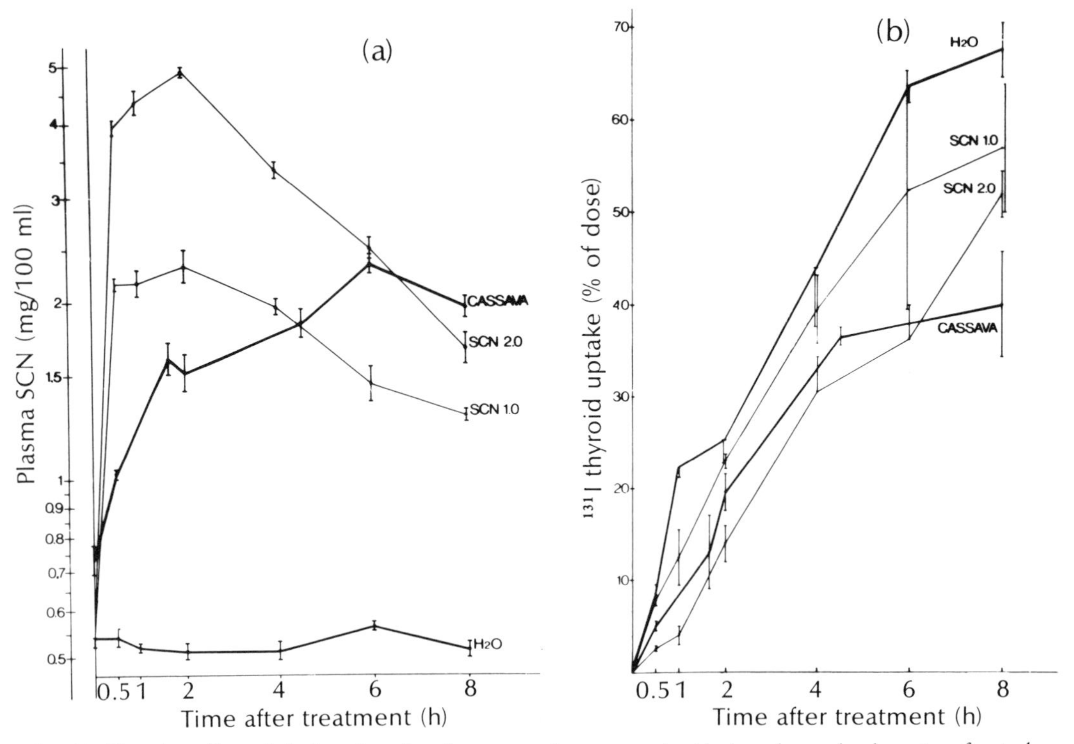

Fig. 56. Short-time effects of the ingestion of crude cassava tubers compared with those due to the absorption of a single thiocyanate dose (1.0 or 2.0 mg) in rats: (a) modifications of SCN plasma concentrations and (b) concomitant ^{131}I thyroid uptake values. Control animals received distilled water; ^{131}I was injected intraperitoneally at time 0.

sively, and after 8 hours the level was higher (22.9 ± 1.5 μg/ml) than those produced by ingestion of 1 mg thiocyanate (Fig. 56a).

In control rats (those receiving distilled water) the plasma thiocyanate concentrations remained unaltered (mean 5.3 ± 0.3 μg/ml).

The ingestion of cassava or thiocyanate (1 and 2 mg) caused an inhibition of thyroid ^{131}I uptake that was closely related to the plasma thiocyanate concentration (Fig. 56b). After 8 hours, this inhibition was most important in group 4 (those receiving cassava) and was accompanied by high plasma thiocyanate concentrations.

Distribution of [^{14}C]CN and [^{14}C]SCN in plasma and erythrocytes: ^{14}C radioactivity, expressed as a percentage of the injected dose, rapidly increased and reached a maximum value after 5 minutes in erythrocytes, in brain, and in the thyroid gland and at 15 minutes in plasma; it then slowly decreased. In contrast, in liver it rapidly decreased during the first 15 minutes and decreased more slowly thereafter. The highest activities were observed in plasma, erythrocytes, and liver. Scanning of thin-layer chromatograms showed that at any one time erythrocytes only contained [^{14}C]CN; plasma contained only [^{14}C]CN during the first 30 minutes. Later, the ^{14}C activity in plasma was also found as thiocyanate: 20% after 30 minutes and 25% after 60 minutes (Fig. 57).

Discussion and conclusion

The effect on thyroid iodine uptake induced by a single meal of cassava in rats is identical to that observed in humans (Delange and Ermans 1971). Moreover, our observations indicated that this inhibition was due to an increase of the plasma thiocyanate concentration, confirming the hypothesis of endogenous thiocyanate production; indeed it should be noted that cassava contains neither thiocyanate nor precursors of thiocyanate but large amounts of linamarin. The time between ingestion of a cassava meal and the onset of an increased plasma thiocyanate level was about 15–30 minutes, probably reflecting the time necessary for endogenous conversion of cyanide into thiocyanate. This conversion appears to be a relatively slow process because 1 hour after ingestion of labeled cyanide, only 25% of plasma radioactivity was found in the form of thiocyanate. However, 2 hours after ingestion of cassava, the [^{14}C]SCN levels were already increased by a factor of two and were still increasing during the next 4 hours. This finding suggests that the yield of the conversion of cyanide is important from a quantitative point of view.

We conclude that ingestion of a single meal of crude tubers of cassava induces a marked rise of the thiocyanate serum concentrations that is associated with a concomitant inhibition of

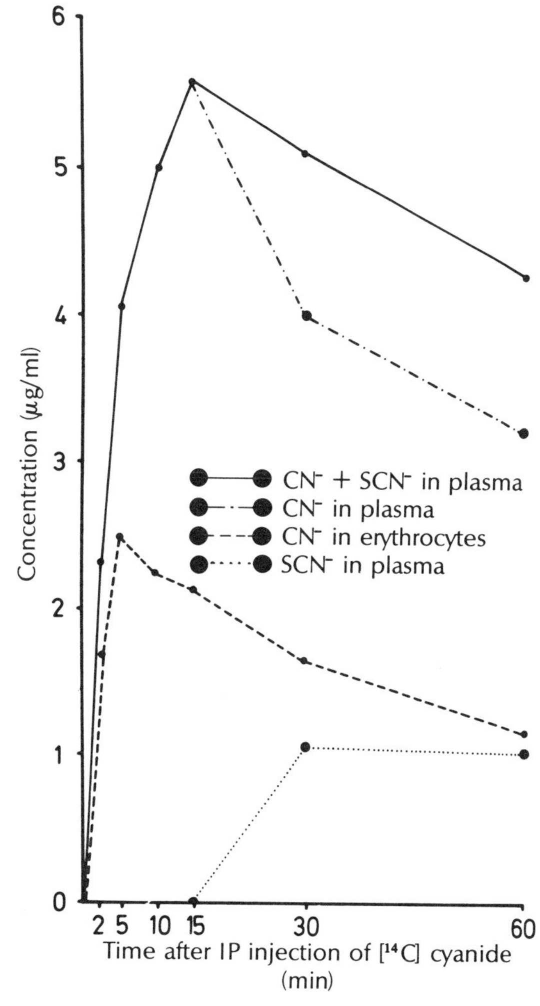

Fig. 57. Modifications of CN^- and of SCN^- concentrations in plasma and in erythrocytes during the hour following the intraperitoneal injection of [^{14}C] cyanide (250 μCi ^{14}C/310 μg CN^-).

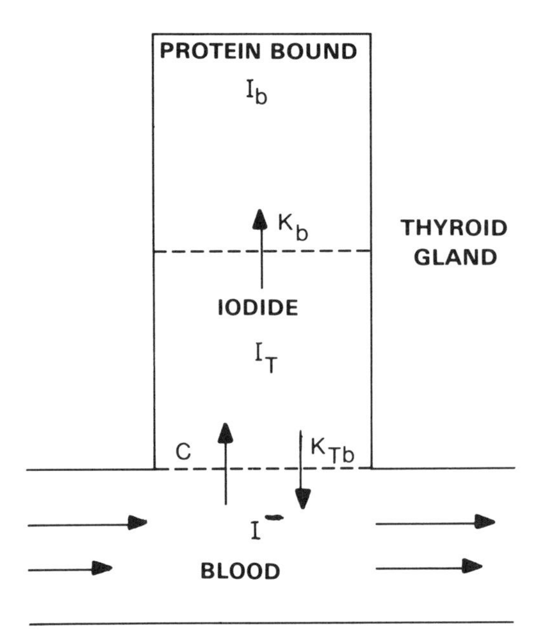

Fig. 58. *Model of iodine kinetics in the thyroid cell according to Wollman (1962); C = unidirectional clearance from plasma iodide into the gland;* K_{Tb} *=: constant-exit rate of thyroidal iodide into the blood;* K_b *= organification constant rate of intrathyroidal iodide;* I_T *= intrathyroidal iodide pool; and* I_b *= intrathyroidal protein-bound iodine pool.*

the thyroid uptake of radioactive iodine. Moreover the shape of the serum thiocyanate curves as well as the fate of intraperitoneally injected [^{14}C]cyanide suggested that the detoxification of cyanide into thiocyanate is a rather slow process that takes several hours to reach equilibrium.

Adaptation of the Thyroid Gland to Chronic Thiocyanate Overload

In the preceding chapters, we have emphasized that the administration of large amounts of thiocyanate (SCN^-) to animals and humans induces the development of goitre. A similar result has been observed in animals after ingestion of plant foods containing certain glucosides whose degradation leads to the endogenous production of SCN^-. It has been supposed that such a goitrogenic action would only be operative in humans after ingestion of very large amounts — thus, the conclusion that the goitrogenic plants do not play a real pathogenic role (Yamada et al. 1974).

However, in the endemic goitre in north Zaire, recent data have suggested that increased serum and urinary SCN^- concentrations and iodine deficiency are jointly responsible for the observed abnormalities of thyroid function, i.e., decreased serum T_4 and increased serum TSH concentrations. We have demonstrated that in this population SCN^- overload is linked to cassava ingestion (Bourdoux et al. 1978).

The mechanism of action of SCN^- was not clear previously because SCN^- overload amounted to only approximately 10 mg/day. In addition no abnormality specifically due to SCN^- had been detected. For example, measures of thyroid iodine uptake were similar in subjects with normal and elevated serum SCN^- levels.

It has been found that the antithyroid action of SCN^- is essentially due to inhibition of the thyroid iodide pump. The exact mechanism is complex and has not been completely elucidated (Wollman 1959; Scranton and Halmi 1969). After injection of small amounts (0.5 mg), SCN^- has little or no effect on unidirectional iodine clearance (C in Fig. 58) and acts by increasing the constant of iodine efflux from the thyroid (K_{Tb}). For a 10-fold higher dose, C is partly or totally inhibited and K_{Tb} is clearly elevated. At these higher concentrations it is probable that SCN^- also inhibits iodine organification and therefore reduces the constant K_b (Fig. 58).

It was recently shown that SCN^- can also modify the interaction of thyroid hormones with their serum-binding proteins (Langer et al. 1976).

A majority of all the earlier studies involved single injections of SCN^-, and little data exist on prolonged SCN^- supplementation. Thus, the purpose of our work was to study the accumulation of iodide and the iodide pump in iodine-deficient rats receiving small daily supplements of SCN^- (0.25 mg/day). The experimental system was designed to mimic the clinical situation observed in the endemic goitre of north Zaire. Thyroid uptake was measured simultaneously with ^{125}I and [^{99m}Tc]-pertechnetate. Pertechnetate has properties similar to iodine but is not organified. We used these two tracers in an attempt to quantify iodine transport whether or not the organification mechanism contributed. In a second set of experiments, we performed our evaluations following interruption of chronic SCN^- supplementation. Finally, we compared the influence of increasing doses of SCN^-, administered intraperitoneally, on thyroid uptake of ^{125}I and ^{99m}Tc.

Results

Thyroid function in rats receiving small daily doses of SCN^-: The mean serum SCN^- concentration, thyroid ^{125}I uptake, and the thyroid/serum ^{99m}Tc ratio (T/S–^{99m}Tc), obtained in four successive experiments (A,B,C, and D) are presented in Table 40.

In each experiment, half of the rats received 0.25 mg SCN^- per day for 20–81 days. All the rats were on a low-iodine diet. Serum SCN^- concentrations were 9–12 μg/ml in the treated animals as compared with 2–4 μg/ml in the controls.

The 4-hour ^{125}I uptake was not inhibited in rats receiving SCN^- supplements. In contrast, the T/S–^{99m}Tc was very significantly reduced in these animals by 67.2, 58.7, and 58.6% in experiments A, B, and C respectively.

The results of other measures of thyroid function showed that, compared with the control rats, the SCN^--supplemented animals demonstrated (Table 40):

- A moderate increase in thyroid weight (mean 12%), which was not statistically significant;
- A highly significant decrease in serum thyroxine in three of the experiments;
- A significant increase of serum TSH; and
- A similar reduction of the thyroid iodine stores.

Table 40. Influence of prolonged supplementation (3–11 weeks) with thiocyanate (0.25 mg/d) in low-iodine diet (LID) rats on the ^{125}I thyroid uptake and the ^{99m}Tc–T/S ratio.[a]

Parameter	Experiment[b]			
	A	B	C	D
Serum thiocyanate (μg/ml)				
LID	1.9 ± 0.5	2.1 ± 0.6	3.7 ± 1.1	3.9 ± 1.3
LID + SCN	12.3 ± 0.6**	8.9 ± 1.5**	11.0 ± 0.8**	11.8 ± 0.8**
4-h thyroid uptake (% of dose)				
LID	27.6 ± 4.8	24.5 ± 3.6	33.1 ± 8.6	76 ± 5
LID + SCN	30.0 ± 3.9	29.2 ± 7.3	38.1 ± 3.0	76 ± 5
^{99m}Tc–T/S ratio				
LID	195 ± 30	177 ± 20	205 ± 64	
LID + SCN	64 ± 18**	62 ± 8**	85 ± 26**	
Serum thyroxine (μg/dl)				
LID	5.2 ± 1.2	4.1 ± 0.7	4.8 ± 1.1	2.2 ± 0.3
LID + SCN	3.1 ± 1.0*	3.9 ± 0.9	3.4 ± 1.0**	1.6 ± 0.5*
Serum TSH (μU/dl)				
LID	10.7 ± 6.0	6.3 ± 4.2		
LID + SCN	11.9 ± 5.4	14.0 ± 8.1*		
Iodine content (μg)				
LID			1.27 ± 0.37	0.60 ± 0.11
LID + SCN			1.14 ± 0.35	0.77 ± 0.31

[a]The mean values ±SD were calculated for 9 rats in experiment A and 20 rats in B, C, and D.
[b]Significance: * = $P < 0.01$; ** = $P < 0.005$; *** = $P < 0.001$.

Comparison of the T/S-^{99m}Tc and T/S-^{125}I ratios: The thyroid/serum ^{99m}Tc ratio was compared with the thyroid/serum ^{125}I ratio in rats in which organification was blocked by propylthiouracil. Half of the animals received 0.25 mg SCN^- per day. The T/S-^{125}I ratio was 30% higher than the T/S-^{99m}Tc ratio in both control and supplemented animals. In supplemented rats, both ratios were markedly reduced (61% for iodide and 66% for ^{99m}Tc).

Effect of discontinuation of SCN^-: After 19 days of low-iodine diet and SCN^- supplementation (0.25 mg/day), SCN^- administration was discontinued (time 0). The T/S-^{99m}Tc ratio and the thyroid ^{125}I uptake at 1 hour were followed daily for 4 days as shown in Table 41. The interruption of thiocyanate intake was followed by a progressive decrease in serum thiocyanate concentration, from 10.5 μg/ml to 5.9 μg/ml after 4 days. During the same lapse of time, we observed a continuous increase in the T/S-^{99m}Tc. This increase was significant at 24 hours, and at 4 days it attained 189% of the initial value (time 0). The ^{125}I uptake showed a similar rise, with a significant increase after 24 hours and a 173% increase at 4 days.

Influence of increasing doses of SCN^- on thyroid uptake of ^{125}I and ^{99m}Tc: Iodine-deficient rats received a single intraperitoneal injection of 0.1–5.0 mg SCN^-. Table 42 shows their thyroid uptake of ^{125}I and ^{99m}Tc 1 hour after the administration of tracers. The results have been expressed as a percentage of uptake observed in control rats who received no SCN^-. For doses less than 1.0 mg, the inhibition of ^{125}I uptake was slight (approximately 15%) and less marked than the inhibition of ^{99m}Tc uptake. For doses greater than 5 mg, SCN^- uptake of both tracers was completely inhibited.

Table 42. Comparative inhibition of the thyroid uptake of ^{125}I and ^{99m}Tc after single injection of graded doses of thiocyanate.

SCN injected dose (mg)	^{125}I thyroid uptake	^{99m}Tc thyroid uptake[a]
0	100 ± 12	100 ± 16
0.1	84 ± 12	69 ± 10
0.5	80 ± 6	41 ± 8
1.0	67 ± 7	24 ± 2
5.0	12 ± 3	6 ± 1

[a]Uptake values are expressed as a percent of values observed in rats receiving no SCN. Measures were carried out 1 h after simultaneous injection of both tracers. Thiocyanate was injected 45 min before the tracers. Each value corresponds to the mean ±SD of the data observed in six rats.

Fig. 59 shows the linear relation between the T/S-^{99m}Tc ratio and the serum SCN^- concentration plotted on a logarithmic scale. The serum SCN^- concentration producing 50% inhibition was of the order of 10 μg/ml.

Discussion and conclusion

The prolonged administration of moderate doses of SCN^- (0.25 mg/day) to iodine-defi-

Table 41. Short-term modifications of ^{99m}Tc-T/S ratio and of ^{125}I thyroid uptake following discontinuation after 19 days of SCN supplementation.

Parameter	Days after SCN^- withdrawal[b]				
	0[a]	1	2	3	4
Serum thiocyanate (μg/ml) (mean ±SE)	10.5 ± 1.1	8.0 ± 1.0**	7.4 ± 0.5**	7.0 ± 0.7**	5.9 ± 0.4**
$^{99m}TcO_4$-T/S ratio (mean ±SE)	69 ± 10	89 ± 15*	99 ± 19**	108 ± 23**	125 ± 17**
1-h ^{125}I thyroid uptake (% of dose) (mean ±SE)	25.1 ± 4.2	32.9 ± 5.5*	29.2 ± 5.8	35.9 ± 5.8**	43.5 ± 7.0**

[a]0 = last day of supplementation.
[b]Significance: * = $P < 0.01$; ** = $P < 0.005$; *** = $P < 0.001$.

cient rats satisfactorily reproduced the abnormalities found in subjects presenting elevated serum SCN^- concentrations in the endemic goitre region of Ubangi. The serum concentrations were alike, 10 μg/ml, the serum TSH increased, and the serum T_4 reduced. Despite the elevated serum SCN^- levels, the thyroid ^{125}I uptake was not diminished.

Our study also showed a marked reduction of the T/S-^{99m}Tc ratio during chronic SCN^- administration to 50–60% of the values in the iodine-deficient controls. This unexpected observation was not due to the pertechnetate itself as it was also observed for ^{125}I in thyroids blocked by propylthiouracil. Thus, despite a normal ^{125}I uptake, low concentrations of SCN^- did alter the iodide pump. A similar inhibition of T/S-^{99m}Tc ratio was observed after intraperitoneal administration of increasing doses of SCN^-.

Under these conditions, the inhibition was strictly proportional to the serum SCN^- concentration. The degree of inhibition observed during chronic SCN administration was similar to the one observed after single-dose administration for corresponding SCN^- serum concentration.

It may be asked whether the difference between the ^{99m}Tc and ^{131}I uptake may be explained solely on the basis of the absence of organification of pertechnetate and whether our results reflect a quantitative alteration in the mechanism of iodide transport.

A number of authors have suggested that low doses of SCN^- only affect the iodide efflux constant (K_{Tb}) without modifying C or K_b. According to this view, iodide, which penetrates the gland, is immediately organified and the intrathyroidal iodide pool remains negligible. In this case, even if K_{Tb} is greatly increased by SCN^-, the fraction of ^{125}I that reexits from the gland is minimal and ^{125}I uptake remains unchanged. This mechanism has been invoked by Wollman and Reed (1958) to explain the difference in SCN^- action on thyroid iodide transport in the presence and absence of propylthiouracil. This also explains the difference in ^{125}I and ^{99m}Tc uptake after large single doses of SCN^- (Fig. 59).

According to this hypothesis, the constant C should not change during thiocyanate

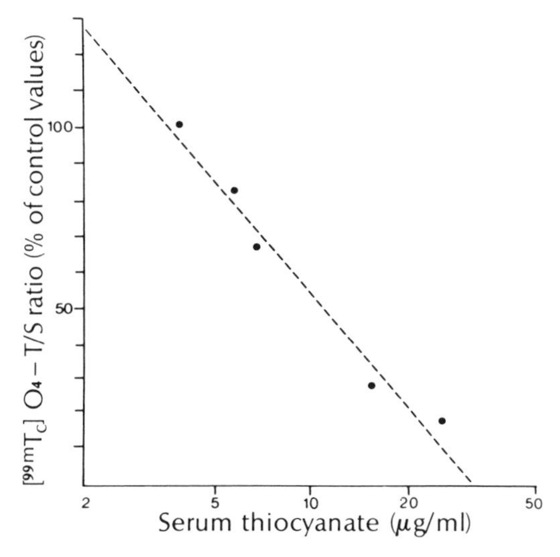

Fig. 59. Relation between the values of [^{99m}Tc] T/S ratio (expressed as % of controls) with the SCN levels in plasma, after single intraperitoneal injection of graded doses of thiocyanate (0.1 to 5 mg).

overload. Our study demonstrated that this is not so. The discontinuation of SCN^- supplementation was followed by a rapid increase in ^{125}I uptake, which paralleled the increase in T/S-^{99m}Tc. The rise in ^{125}I uptake reflected an increase in unidirectional iodide clearance that was masked in the presence of SCN^-. Therefore, the maintenance of thyroid ^{125}I uptake in rats receiving daily doses of SCN^- was the result of an adaptation in which an increase in unidirectional clearance compensated the thiocyanate-induced abnormalities in iodide transport. The measurement of ^{99m}Tc uptake revealed the increased iodide reexit, which was already evident for very slight increases in serum SCN^- concentration.

Our studies did not exclude the possibility that, under our experimental conditions, SCN^- also acts at other sites, specifically on organification.

In conclusion, the prolonged administration of low doses of SCN^- to iodine-deficient rats has important effects on the iodide pump, in particular on iodide reexit. These effects cannot be detected by conventional studies using radioactive iodine because of a compensatory increase in unidirectional clearance. Our findings may explain the antithyroid action of certain plant foods, even though the amounts of

Table 43. Chronic action of manganese ($MnCl_2$) on iodine metabolism of low-iodine diet (LID) rats.

Experiment	Diet	Drugs added to the diet: SCN^- (mg)	Drugs added to the diet: Mn^{++} (μg)	Sample	Thyroid weight (mg/100 g)[a] (mean ± SE)	6-h ^{131}I thyroid uptake (% of dose)[a] (mean ± SE)	6-h plasma $PB^{131}I$ (% of dose/ml)[a] (mean ± SE)	11-d ^{125}I thyroid uptake (% of dose)[a] (mean ± SE)	11-d plasma $PB^{125}I$ (% of dose/ml)[a] (mean ± SE)
1	LID	5	0	8	17.8 ± 1.8	63.9 ± 5.3	0.096 ± 0.028	6.2 ± 1.4	0.073 ± 0.015
	(48 d)	5	25	6	14.4 ± 1.5**	58.7 ± 3.9	0.052 ± 0.022*	10.5 ± 2.3**	0.063 ± 0.006
		5	50	6	14.8 ± 1.7**	53.5 ± 4.9**	0.062 ± 0.016**	12.3 ± 2.2**	0.063 ± 0.004
2	LID	5	0	9	21.4 ± 2.6	56.2 ± 4.2	0.324 ± 0.116	5.6 ± 3.5	0.059 ± 0.017
	(35 d)	5	50	9	22.3 ± 5.1	55.8 ± 7.8	0.428 ± 0.105	4.4 ± 2.4	0.039 ± 0.014*
3	LID	0	0	9	11.2 ± 1.7	32.8 ± 4.8	0.09 ± 0.13	12.1 ± 4.8	0.05 ± 0.02
	(46 d)		50	9	9.2 ± 1.5	24.8 ± 5.7*	0.06 ± 0.15	11.0 ± 3.2	0.03 ± 0.01
4	LID	0	0	6	14.7 ± 1.6	37.4 ± 4.5	0.03 ± 0.01	35.4 ± 8.7	0.06 ± 0.12
	(46 d)		50	6	15.8 ± 1.8	44.4 ± 6.8	0.02 ± 0.05	45.6 ± 7.4	0.10 ± 0.02
5	LID +5 μg I^-/d (43 d)	0	0	8	6.4 ± 0.09	8.0 ± 3.2	0.02 ± 0.01	5.5 ± 1.8	0.02 ± 0.01
			50	8	6.3 ± 1.0	8.2 ± 1.2	0.02 ± 0.01	5.3 ± 2.0	0.02 ± 0.01

[a]Significance: * = $P < 0.01$; ** = $P < 0.005$.

thiocyanate they provide are very small compared with the amounts thought to be necessary for a goitrogenic effect in humans.

Influence of Manganese

A study of chemical composition and oligo-elements content on Idjwi Island was undertaken by Cornil et al. (1974). These authors compared the soil and plants of the goitrous region (north) with those of the nongoitrous region (south). Their most striking finding was a reduction in manganese content in plants and, in particular, in cassava cultivated in the northern part of the island.

McCarrison and Moldhava showed in 1923 that the administration of manganese chloride to rabbits fed solely on cabbage had a clear, antigoitrogenic effect. More recently Kaelis (1970) reported that administration of manganese chloride to rats facilitated intrathyroidal iodide metabolism, and, in particular, thyroid hormone secretion. Manganese ion has been shown under certain in vitro conditions to activate the iodination of tyrosine by peroxidase (De Groot et al. 1968).

On the basis of these data, we have attempted to test the hypothesis that the epidemiological difference between the north and southwest of Idjwi was linked to a relative manganese deficiency in the goitrous region. In the present investigation, we have studied the effect of manganese on various measures of thyroid function in the rat, in the presence or absence of iodine deficiency and/or SCN^- supplementation.

Methods

In most of the experiments, the effect of manganese was tested during chronic administration of 25–50 μg $MnCl_2$ per day, added to a Remington diet to which potassium iodide (5 μg/day) or SCN^- (5 mg/day) was added when appropriate. After 3–6 weeks the rats were killed and exsanguinated. For evaluation of thyroid metabolism, ^{125}I and ^{131}I were injected, respectively, 11 days and 6 hours before sacrifice.

In two experiments, the effects of a single intraperitoneal injection of 100 or 250 μg $MnCl_2$ were investigated in iodine-deficient rats.

Results

Table 43 shows the results obtained in five separate experiments on rats receiving an iodine-poor diet (numbers 1–4), an iodine-rich diet (number 5), and SCN supplement (numbers 1 and 2). In the first experiment (1), rats given 25 and 50 μg $MnCl_2$ per day had diverse modifications in thyroid function, i.e., decreased thyroid weight, ^{131}I uptake, and $PB^{131}I$ (6th hour). In addition there was a clear increase in the retention of ^{125}I, administered 11 days previously.

None of these effects was observed in the second experiment (2), which was identical to the first, nor was there any effect in experiments 3 and 4 involving iodine-deficient rats, although in the third experiment (3) there was a reduction in 6-hour ^{131}I uptake. In rats on an iodine-rich diet (5) there was no difference between controls and animals receiving manganese.

Tables 44 and 45 show the results of a single intraperitoneal injection of $MnCl_2$ (100 and 250 μg) on 24-hour ^{131}I uptake, T/S-^{99m}Tc ratio, and serum T_4 and TSH concentrations. No significant modification was observed.

Complementary studies were performed to test the hypothesis that manganese can modify the conversion of thyroxine to triiodothyronine. Thyroxine labeled with ^{125}I was injected into iodine-deficient rats with or without manganese supplementation (100 μg/day). The half-life of the labeled T_4 proved to be identical in the two groups of animals.

Discussion and conclusion

A clear antigoitrogenic effect of manganese was observed in one experiment (1) but could

Table 44. Effects of intraperitoneally injecting 12 LID rats with a single 100 μg dose of Mn^{++} ($MnCl_2$) compared with 12 controls at 24 h post-injection.

Parameter	Control rats	Manganese-treated rats	P value
Captation, 24-h ^{131}I uptake (mean ±SE)	40.6 ± 8.0	37.0 ± 12.9	NS
Serum $PB^{131}I$/ml (% of dose)[a] (mean ±SE)	0.272 ± 0.101	0.391 ± 0.119	NS
Serum $PB^{125}I$/ml (% of dose)[a] (mean ±SE)	0.075 ± 0.019	0.085 ± 0.023	NS

[a] ^{131}I and ^{125}I were injected respectively 1 and 15 days before sacrifice.

Table 45. Effects of intraperitoneally injecting 22 LID rats with a single 250 μg dose of Mn^{++} ($MnCl_2$) compared with 22 controls at 4½ h.

Parameter	Control rats	Manganese-treated rats	P value
^{99m}Tc–T/S ratio (mean ±SE)	113 ± 21	130 ± 29	< 0.01
Serum $PB^{125}I$/ml (% of dose) (mean ±SE)	0.42 ± 0.11	0.50 ± 0.17	NS
Serum T_4 (μg/dl) (mean ±SE)	4.0 ± 1.0	3.5 ± 0.9	NS
Serum TSH (μU/ml) (mean ±SE)	4.4 ± 1.6	4.7 ± 1.8	NS

not be reproduced in subsequent experiments. The discrepancy was all the more difficult to interpret as the indices of thyroid function in the first experiment were significantly modified and in a consistent manner, suggesting a reduction in thyroid iodide turnover and thyroid stimulation. These results might be due to an accidental contamination with iodide in the manganese-treated rats.

The present observations did not reveal an action of manganese on thyroid function. However, the possibility that manganese influences thyroid metabolism in iodine-deficient rats under other conditions cannot be excluded. Recently, Buthieau and Autissier (1977) have reported that a dose of 2 mg/100 g manganese in the rat produced an important decrease in thyroid uptake and serum T_4 and T_3 concentrations. In view of the enormous quantities of Mn employed, it seems unlikely that these results have pathophysiologic importance.

Chapter 9

Fetomaternal Relationship, Fetal Hypothyroidism, and Psychomotor Retardation

C.H. THILLY, G. ROGER, R. LAGASSE, D. TSHIBANGU, J.B. VANDERPAS, H. BERQUIST, G. NELSON, A.M. ERMANS, AND F. DELANGE

The major complication of severe endemic goitre is cretinism. The cause is unknown, but it has been proposed that cretinism is linked to hypothyroidism during fetal life or shortly after birth (Dumont et al. 1963b; Delange and Ermans 1971b). This hypothesis has never been proved. It arises from indirect evidence. First, there is an obvious clinical similarity between endemic and sporadic cretinism observed in industrialized countries. In sporadic cretinism, hypothyroidism at birth is well demonstrated (Andersen 1961; Maenpaa 1972) and correction of thyroid insufficiency in the first 3 months of life is essential to avoid the major complication of the disease — mental deficiency (Smith et al. 1957; Raiti et al. 1971; Klein et al. 1972; Dussault et al. 1978). In addition, the radiologic changes of endemic cretinism, such as epiphyseal dysgenesis, are also pathognomonic of the hypothyroidism of fetal life or the newborn period (Wilkins 1941; Dumont et al. 1963a; Delange et al. 1972c). Finally, the observation (Pharoah et al. 1971, 1972; Ramirez et al. 1972; Thilly et al. 1973a) that the introduction of prophylaxis with iodine in endemic regions prevents cretinism is another argument in favour of the etiologic role of hypothyroidism, considering the essential role of iodine in thyroid hormogenesis.

An important question, which has not been well studied in regions of severe endemic goitre is how does thyroid function in the mother relate to that in the fetus. In nonendemic regions, no correlation exists between hormone levels in the newborn and the mother (Carr et al. 1959; Fisher et al. 1969, 1970), but it is essential to know whether this is true in regions where both iodine deficiency and thiocyanate affect the thyroid function of the mother and the fetus. It is known that these two ions cross the placenta freely.

Another question not yet answered is whether only frank cretins in endemic regions are mentally deficient or whether there is a retardation in psychomotor development in all children in these regions (Stanbury 1974). Preliminary studies have been carried out in regions of endemic goitre in Peru (Pretell et al. 1972) and Ecuador (Dodge 1969; Trowbridge 1972; Fierro-Benitez et al. 1972, 1974a). In Peru, the authors observed a slight increase in the intellectual quotient (IQ) of children aged 1–58 months, born from mothers treated with iodized oil, but the IQ was not significantly different from that of a comparable group of children of untreated mothers. In Ecuador, significant increases in IQ were observed in groups of older children (3½–9 years) of mothers treated with iodized oil. However, these children were from different villages with different sociocultural levels and nutritional patterns and where the cretinism described, essentially neurologic, was very different from that observed in Zaire. Thus, it is not known whether there is a deficiency in psychomotor and intellectual development in the general population of these regions. This is an extremely important question from the point of view of public health, as widespread intellectual impairment could be a powerful impediment to the socioeconomic development of entire regions.

The purpose of our study, therefore, was to investigate:

- The existence of congenital hypothyroidism in the newborn;
- The possible relations between thyroid function in the mother and that in the newborn; and
- The existence of a delay in psychomotor development in regions of severe endemic goitre.

Therapeutic and Prophylactic Trial

A therapeutic and prophylactic trial was carried out in the Karawa hospital, which serves a rural population of approximately 30 000 people. The hospital is located in the centre of the goitrous region of Ubangi, where the prevalence of goitre is high and homogeneous (average 62%) and cretinism is present (average 4.7%). Fig. 60 shows the general plan of the trial. One group of pregnant women received an injection of iodized oil, containing 475 mg slowly absorbable iodine in a volume of 1 ml, at the first prenatal visit. The control group received an injection of vitamins, free of iodine. The first prenatal visit was during the second or third trimester, on the average at the 28th week. The two groups were constituted at random using random number tables and numbered envelopes. The two groups were strictly comparable, possessing the same characteristics of age, parity, living conditions, dietary habits, etc. Table 46 shows the absence of significant differences between the groups

Table 46. Comparison of the two groups of women assigned at random to the treated or untreated group during the first prenatal visit.[a]

Parameter	Untreated (332)[b]	Treated (339)
Mean age (y)	22.8 ± 5.9	23.4 ± 6.3 NS
Goitre prevalence (%)	70.6	70.2 NS
Number of previous pregnancies	2.9 ± 2.3	3.0 ± 2.3 NS
Uterine height (cm)	21.0 ± 3.0	21.0 ± 3.5 NS

[a]Significance: NS = not significant; mean ± SD.
[b]Figure in parentheses is number of women.

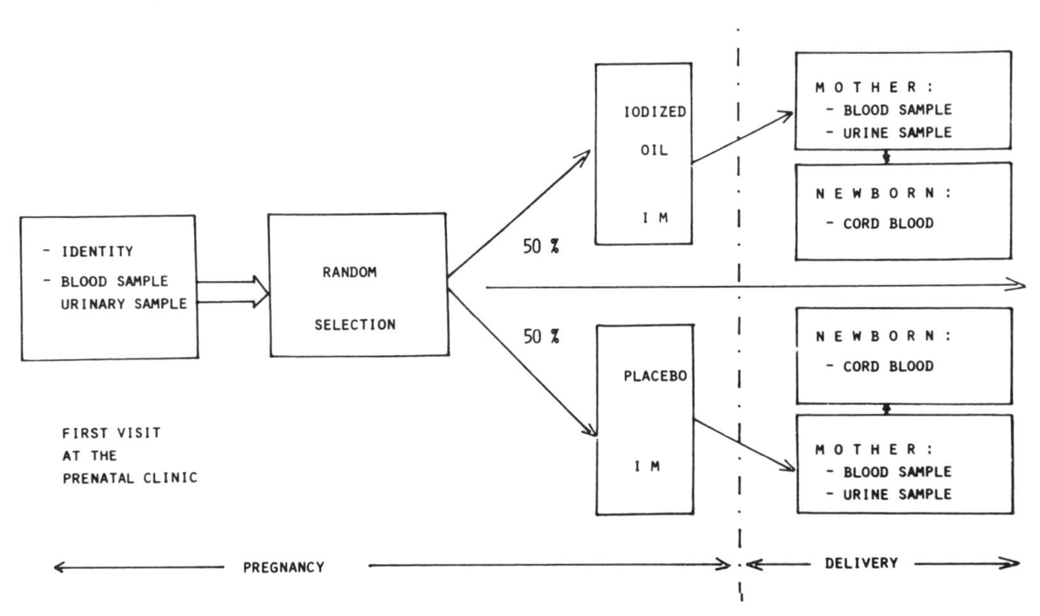

Fig. 60. General plan of therapeutic and prophylactic trial of iodized oil administered during pregnancy.

of 339 treated and 332 untreated women. Immediately before the administration of iodized oil, urine and blood samples were taken for measurement of urinary iodine and serum T_3, T_4, and TSH. Sixty percent of the women in the study gave birth in the hospital. At delivery, blood was drawn from the mother and from the umbilical cord of the baby for determination of serum hormone levels.

The study was a prospective and open investigation. The first women were enrolled in February 1973 and the rest, at 10 per month until 1977. We recalled all the children to compare the psychomotor development of offspring of treated and nontreated mothers, spreading the examinations over 4 years. In August 1973, January 1974, April 1974, November 1974, January 1975, November 1975, August 1976, and August 1977, groups of children (aged 4 months to 3 years) were evaluated. We performed all interviews and examinations under strict double-blind precautions, i.e., neither the mother nor the investigator knew whether the mother had received the iodine supplement. The investigators evaluated psychomotor development according to the Brunet-Lézine scale, by measurements of developmental quotient, by comparison of frequency of success in certain tasks, and by comparison of the time of acquisition of four specific items of the scale.

Thyroid Status of the Participants

During pregnancy: Fig. 61 shows the results of measurement of serum T_4, T_3, and TSH in 162 untreated pregnant women as a function of gestational age. The normal range is also indicated (Fisher 1975). Approximately half of the women presented T_4 values inferior to the limits of normal during pregnancy, and approximately 25% had elevated TSH values (20–100 μU/ml). The T_3 values were more widely dispersed than is normally observed during pregnancy. There was no correlation between T_3 and T_4 values in this group of women (r = 0.01, P > 0.20) (Fig. 62). The average values were T_4 11.3 μg/dl, T_3 206 ng/dl, and TSH 8.2 μU/ml. Serum thyroxine-binding globulin (TBG) was measured in 31 untreated women

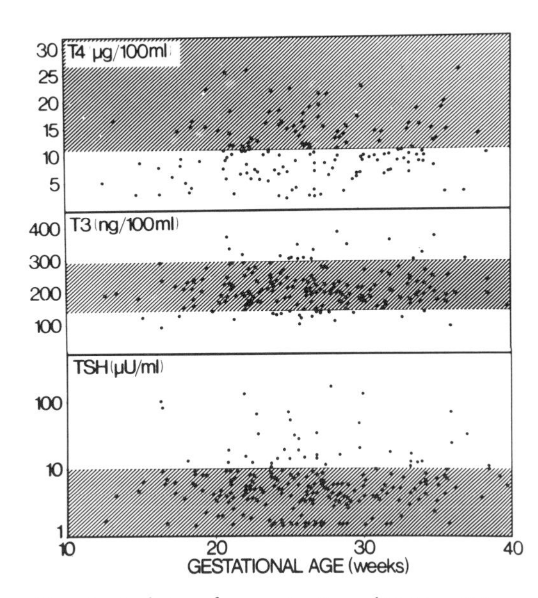

Fig. 61. *Evolution of serum T_4, T_3, and TSH in 122 untreated pregnant women as a function of gestation. The hatched areas are the normal range of maternal values (Fisher and Dussault 1974).*

between the 10th and 40th week. The observed value, 3.54 ± 0.77 mg/dl, was elevated, compared with the value normally observed during pregnancy.

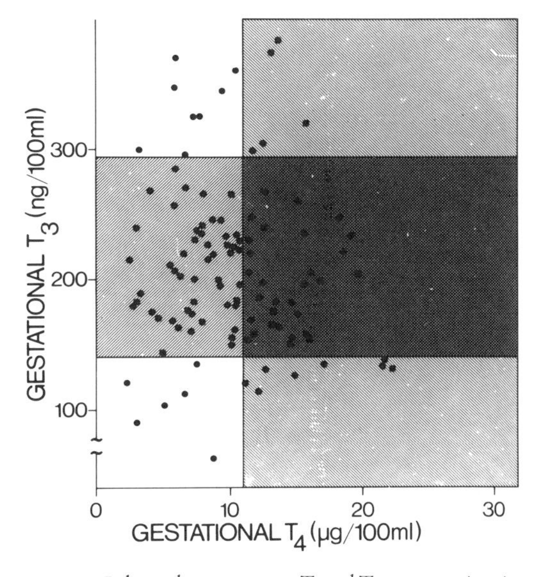

Fig. 62. *Relation between serum T_3 and T_4 concentrations in 108 untreated pregnant women, measured at the first prenatal visit.*

Table 47. Serum T_4, T_3, TBG, and TSH levels in mothers at the time of delivery.

Serum concentration (mean ±SE)[a]	Untreated	Treated
T_4 (μg/dl)	11.5 ± 0.7 (51)[c]	15.7 ± 0.7*** (49)[a]
T_3 (ng/dl)	171 ± 10 (56)	154 ± 9 NS (53)
TBG (mg/dl)	3.02 ± 0.78 (18)	3.02 ± 0.72 NS (16)
TSH (μU/ml)[b]	8.7 (7.6–9.9) (52)	5.4 (4.9–5.9)***(52)

[a]Significance: NS = not significant; * = $P < 0.05$; ** = $P < 0.01$; *** = $P < 0.001$.
[b]Geometric mean ($\bar{G}$ – 1SE, $\bar{G}$ + 1SE).
[c]Number of subjects.

At delivery: Table 47 compares the mean serum concentrations of T_4, T_3, TSH, and TBG at the time of delivery in 56 untreated and 53 treated subjects. The values observed in the untreated group were similar to their values during pregnancy, whereas in mothers receiving treatment the mean serum T_4, 15.7 μg/dl, was more comparable with values observed in Belgian mothers. The mean serum TSH in the untreated group, 8.7 μg/dl, was significantly higher than in the treated group, 5.4 μg/dl, but the mean values for serum T_3 and TBG were not significantly different in the two groups.

The newborn: Table 48 presents the mean concentrations of T_4, T_3, and TSH in serum from cord blood of 60 newborns of untreated mothers and 58 newborns of treated mothers. The serum T_4 concentration, 9.4μg/dl, in newborns of the untreated group was significantly lower than that found in a Belgian control group. The value was normalized to 12.4

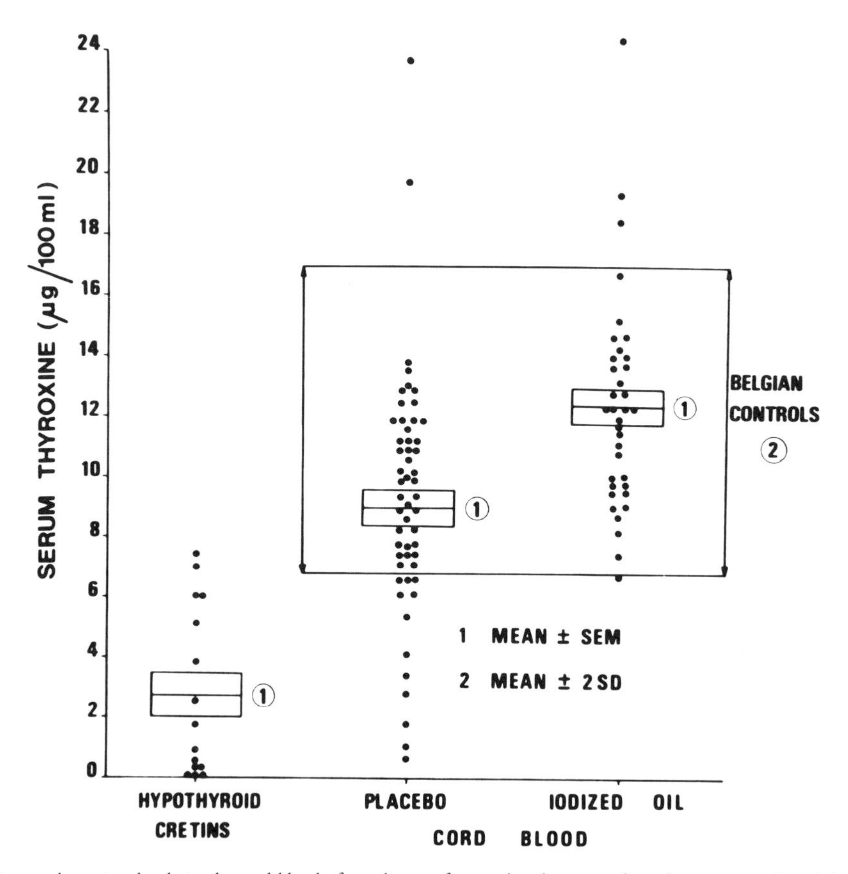

Fig. 63. Serum thyroxine levels in the cord blood of newborns of treated and untreated mothers compared with levels for 15 myxedematous cretins and for newborns in Belgium.

Table 48. Serum T_4, T_3, and TSH levels in newborns of untreated and treated mothers.

Serum concentration (mean ±SE)[a]	Untreated	Treated
T_4 (μg/dl)	9.4 ± 0.8 (34)[c]	12.4 ± 0.5** (47)[a]
T_3 (ng/dl)	68 ± 6 (42)	55 ± 6 NS (45)
TSH (μU/ml)[b]	19.6 (16.6–23.2) (60)	6.4 (5.8–7.0)*** (58)

[a]Significance: NS = not significant; * = $P < 0.05$; ** = $P < 0.01$; *** = $P < 0.001$.
[b]Geometric mean ($\bar{G}$ - 1SE, $\bar{G}$ + 1SE).
[c]Number of subjects.

μg/dl, by treatment of the mother. Serum TSH fell from 19.6 μU/ml in the untreated group to 6.4 μU/ml in the treated group. There was no significant difference in serum T_3 values, 68 vs 55 ng/dl.

Fig. 63 compares the individual values for serum T_4 in the newborns with the values in Belgian newborns and the values in eight myxedematous cretins from Ubangi. Whereas the T_4 values in the treated group could be superimposed on the Belgian values, the serum T_4 values in a third of the newborns of untreated mothers were extremely low and could be superimposed on the values from cretins living in the same region.

Comparison of the thyroid status of the mother and the newborn: The mean values for T_4, T_3, and TSH in untreated mothers and their newborns are illustrated in tables 47 and 48. Compared with their mothers, the newborns presented significantly reduced T_4 (9.4 μg/dl for the newborns vs 11.5 μg/dl for the mothers) and T_3 (68 ng/dl vs 171 ng/dl) values. The serum TSH concentration in the newborns (19.6 μU/ml) was nearly double that observed in the mothers (8.7 μU/ml).

The individual values for TSH in mothers and their newborns are shown in Fig. 64. The means and standard deviations in the treated groups, considered as control populations, are represented by solid horizontal lines. In untreated mothers, only 15% (8 of 52 subjects) had serum TSH concentrations higher than two standard deviations above the mean, whereas 40% (24 of 62) of their newborns exceeded these limits. These findings have demonstrated clearly that in the absence of iodine prophylaxis, serum TSH values in newborns in this goitre endemic area are more profoundly altered than are the values in their mothers.

Six of 34 newborns in the untreated group (18%) had serum T_4 values less than 5 μg/dl. The individual values were 4.7, 3.3, 2.8, 1.9, 1.7, and 1.0 μg/dl. The corresponding T_3 values were 90, 95, 63, 60, unknown, and 38 ng/dl, and TSH values 40, 114, 120, 300, 93, and 300 μU/ml. These values are typical of severe thyroid insufficiency.

Table 49 shows the mean values of thyroid hormones in three groups of untreated mothers and newborns selected on the basis of the maternal serum T_4 concentration: group A, $T_4 > 10$ μg/dl; group B, 10 μg/dl $> T_4 > 8$ μg/dl; group C, $T_4 < 8$ μg/dl. In both mothers and

Table 49. T_4, T_3, and TSH levels in three groups of mothers and their newborns grouped arbitrarily according to the maternal T_4 value (number of subjects in parentheses).[a]

Maternal serum level of T_4 (μg/dl)	Maternal serum at delivery (mean ± SE)			Newborn cord serum (mean ± SE)		
	T_4 (μg/dl)	T_3 (ng/dl)	TSH (μU/ml)[b]	T_4 (μg/dl)	T_3 (ng/dl)	TSH (μU/ml)[b]
(A) $T_4 > 10$	14.8 ± 0.6 (31)	174 ± 11 (27)	6.2 (5.5–7.0) (25)	11.6 ± 1.1 (13)	56 ± 7 (20)	9.4 (7.6–11.6) (22)
(B) $8 < T_4 < 10$	*8.8* ± 0.2***(9)	229 ± 30NS (8)	8.9 (7.2–11.0) NS(8)	7.8 ± 0.7** (7)	75 ± 8NS (6)	22.9 (17.8–29.5)* (8)
(C) $T_4 < 8$	4.5 ± 0.6*** (11)	129 ± 23*** (10)	16.6 (10.3–26.4**) (10)	5.5 ± 16** (7)	68 ± 11 (8)	143.7 (98.2–210.2)** (6)

[a]Significance: NS = not significant; *= $P < 0.05$; ** = $P < 0.01$; ***= $P < 0.001$; the levels indicate relations between groups A and B and groups A and C.

[b]Geometric mean ($\bar{G}$–SE, $\bar{G}$ + SE).

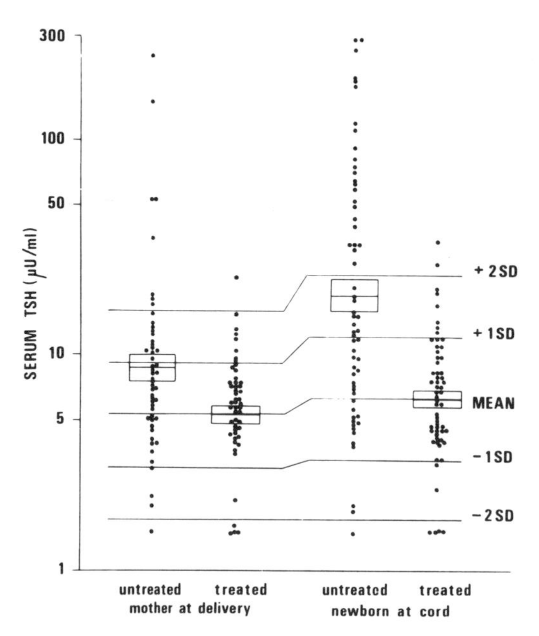

Fig. 64. Serum TSH levels in groups of mothers and newborns of mothers treated and untreated during pregnancy. Standard deviations from the mean value for controls (treated group) are indicated. The boxes indicate the geometric mean ± SE.

newborns serum TSH increased with decreasing T_4. In addition, the low maternal T_4 values (group B and C) corresponded to the lowest T_4 and highest TSH values in the newborns. This finding indicates the existence of a correlation between the maternal and newborn T_4 and TSH values.

In group C, 11 of 51 mothers had serum T_4 values lower than 8 μg/dl, and their mean TSH was 16.6 μU/ml. The newborns of these mothers had a very low mean T_4, 5.5 μg/dl, and an extremely elevated TSH, 144 μU/ml.

Fig. 65 presents TSH values in mothers and newborns as a function of serum T_4 concentration levels at the first prenatal visit (on average, the 24th week). Two groups were constituted, one with T_4 values higher than 8 μg/dl and the other with lower values. The former group had a practically normal mean serum TSH, 5.5 μU/ml. This value was maintained at 5.6 μU/ml at delivery. The newborns of these mothers presented a mean TSH of 11.2 μU/ml, which is not significantly different from the value in Belgian newborns. In contrast, in mothers with prenatal T_4 levels lower than 8 μg/dl, prenatal TSH was 11.2 μU/ml; the value at delivery, 10.7 μU/ml; and the corresponding value for newborns, an extremely elevated, 47.9 μU/ml.

Correlation studies: The correlations between the serum T_4, T_3, and TSH values in the untreated group are presented in Table 50. Highly significant inverse correlations were obtained between T_4 and TSH values both in mothers and in newborns; however there was no correlation between serum T_3 and T_4 or between T_3 and TSH. Significant positive correlations existed between maternal and newborn T_4 levels and TSH values, and there was a significant inverse correlation between maternal T_4 and newborn TSH, and between maternal TSH and newborn T_4 (Fig. 66).

Psychomotor development: We measured the development quotients in three age groups (4–9 months, 10–15 months, 16–23 months), using the Brunet-Lézine scale (Table 51). The quotients of the untreated group were consistently inferior to those of the treated group, and the difference was significant for the 4–9-month age group and for all the children taken together. In addition, for both the treated and untreated groups, the quotient was highest in the youngest age group, decreasing in the older groups.

Table 52 presents the analysis by task in the three age groups; it shows the number of tasks

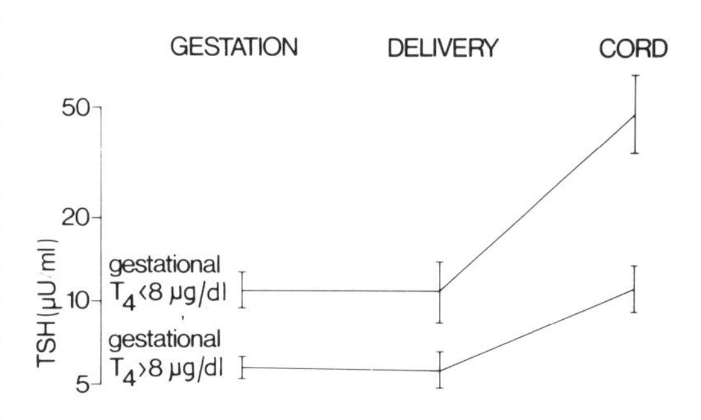

Fig. 65. Comparison of serum TSH concentrations in mothers during pregnancy and at delivery and in newborns (cord blood) in two groups selected on the basis of maternal serum T_4 concentration at the first prenatal visit (higher or lower than 8 μg/dl).

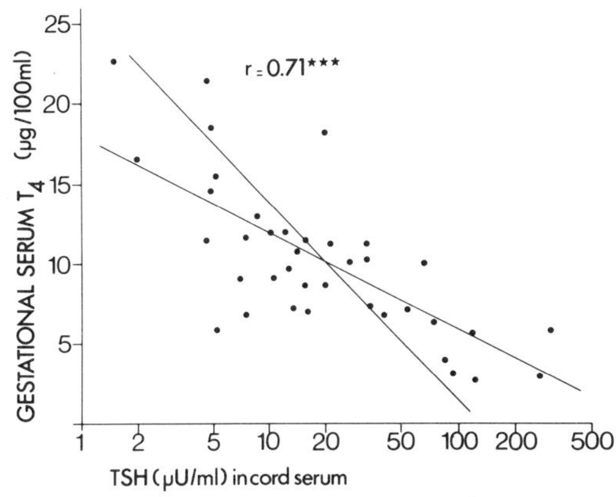

Fig. 66. *Inverse correlation between maternal serum T_4 concentration during pregnancy and serum TSH value in the newborn: correlation coefficient (r) and linear regression.*

more frequently accomplished by the treated group, those more frequently accomplished by the untreated group, and those with an equal frequency of success. In the 4–9-month age group 50% of the tasks were more frequently performed by the treated group, 18% more frequently by the untreated group, and 32% with an equal frequency. The percentage difference was significant ($P<0.001$). For the 10–15-month age group, 42% of the tasks were better accomplished by infants of treated mothers compared with 19% for the untreated (39% with equal frequency). This difference was also significant ($P<0.01$); however, there was no significant difference in the results for the 16–23-month group.

Fig. 67 shows the results for four specific

Table 50. Correlation coefficients between T_4, T_3, and TSH serum levels of mothers and newborns at delivery in the Ubangi endemic goitre region (upper left contains values for mother; lower right, newborns; upper right, mothers and newborns).[a]

	Maternal serum						Newborn cord serum					
	T_4		T_3		TSH		T_4		T_3		TSH	
	n	r	n	r	n	r	n	r	n	r	n	r
Maternal serum												
T_4	51	1	44	+0.17NS	42	−0.47**	26	0.80***	33	−0.15NS	35	−0.79***
T_3			56	1	46	+0.07NS	23	−0.16NS	36	0.21NS	37	0.07NS
TSH					52	1	26	−0.57**	31	0.09NS	34	0.61***
Newborn cord serum												
T_4							34	1	24	−0.29NS	30	−0.72***
T_3									42	1	36	0.30NS
TSH											60	1

[a]n = number of pairs of values; r = correlation coefficient; significance: NS = not significant; * = $P < 0.05$; ** = $P < 0.01$; *** = $P < 0.001$

Table 51. Comparison of developmental quotients in children born from untreated and treated mothers.[a]

		Developmental quotients				
		Untreated		Treated		
Group	Age (months)	number	mean ± SE	number	mean ± SE	*t-test*
I	4–9	14	109 ± 6	15	127 ± 4	2.47*
II	10–15	16	101 ± 4	13	109 ± 5	1.17NS
III	16–23	6	91 ± 12	11	106 ± 4	1.51NS
I–III	4–23	36	103 ± 4	39	115 ± 3	2.70**

[a]Significance: NS = not significant; * = $P < 0.02$; ** = $P < 0.01$.

Table 52. Comparison of the successes by untreated and treated children in performing tasks from the Brunet-Lézine scale.[a]

Group	Age (months)	Children	Tasks performed successfully: More often by untreated	More often by treated	Equally by both	χ^2
I	4–9	44	11	31	20	19.1***
II	10–15	45	12	26	24	10.3**
III	16–23	38	11	16	35	1.9NS

[a]Significance: NS = not significant; * = $P < 0.05$; ** = $P < 0.01$; *** = $P < 0.001$.

items of the Brunet-Lézine scale. The percentage of children accomplishing the task (ordinate axis) is represented as a function of age (abscissa). The characteristic age at which children in an occidental population accomplish this task is also shown, as is the age at which two-thirds of the infants of treated and untreated mothers succeeded. The infants of

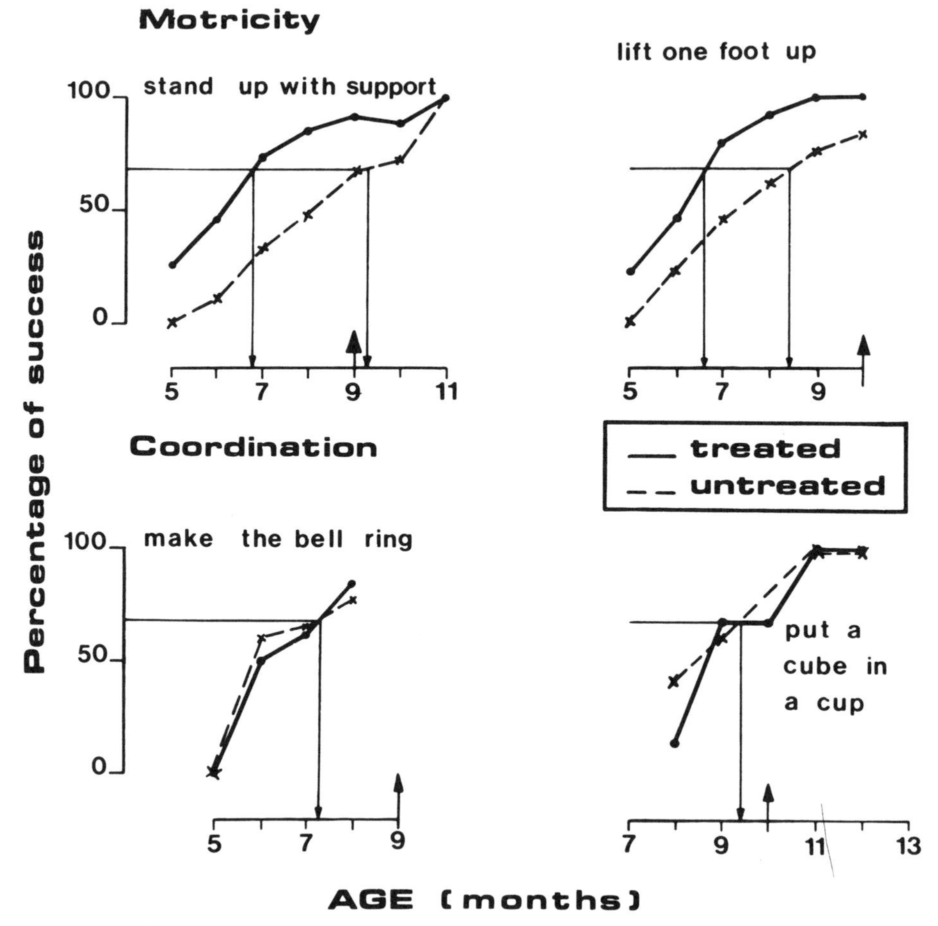

Fig. 67. Percentage of success for four individual items of the Brunet-Lézine scale in relation to the age of the children. The characteristic age at which children in an occidental population accomplish the tasks is shown by an arrow.

the treated group showed superior results in motor development for a wide age range — for example, two-thirds of them could stand with support at 7 months, whereas two-thirds of the untreated group achieved the same task at approximately 9 months. In contrast, no difference between the treated and untreated groups was noted in two tasks measuring coordination — ringing a bell and putting a block in a cup without letting it go.

Discussion and Conclusions

Congenital hypothyroidism in severe endemic goitre: The first original observation of this study was the demonstration that, in a region of severe endemic goitre, an important fraction of newborns present clear biologic signs of congenital hypothyroidism. Approximately a quarter to one-third of the subjects had a serum T_4 concentration less than 6 μg/dl and a TSH higher than 20 μU/ml, which are the normal limits for Belgian newborns. The degree of abnormality of the observed values compared with that obtained in systematic screening for overt congenital hypothyroidism. Thus, our results demonstrated, for the first time, congenital hypothyroidism in severe endemic goitre. By systematic screening in Europe, the United States, and Canada, investigators have calculated an incidence of congenital hypothyroidism of approximately 1 in 5000 births (Klein et al. 1974; Dussault et al. 1975; Buist et al. 1975; Delange et al. 1979; Illig 1979). In Ubangi, the incidence is of the order of 20% and therefore represents a priority public health problem.

Our results have shown that the extreme abnormalities of thyroid function frequently observed in Ubangi newborns are accompanied by a systematic decrease in the mean and overall distribution of T_4 values and by a systematic increase in TSH concentration levels, a process indicating an important reduction in thyroid function throughout the newborn population in Ubangi. These data bring a new and crucial element to the understanding of the pathogenesis of cretinism and mental deficiency in a region of severe endemic goitre.

Thyroid status in the pregnant woman: The characterization of the thyroid status in women during pregnancy was more difficult than in the newborns because we did not have values for the treated groups throughout the period. Nevertheless, the observation in Ubangi that the elevation in T_4 is less than in a normal pregnancy was confirmed by the mean T_4 at delivery, which in the untreated group was very significantly lower than in the treated group. The elevated TBG levels in a limited number of women taken together with several very low total T_4 values leaves no doubt that at least a certain proportion of these women suffered from hypothyroidism during their pregnancy. Pretell et al. (1974a, b) in Peru were the first to demonstrate this situation in a region of endemic goitre; several authors (Greenman et al. 1962; Echt and Doss 1963; Man et al. 1969; Jones et al. 1969) have shown that women with moderate hypothyroidism are at increased risk for abortion, stillbirth, and congenital malformation, independent of iodine deficiency. Thus, in a region of endemic goitre, one must wonder whether the dramatic consequences of congenital hypothyroidism in the child are not accompanied by another set of anomalies reflecting the influence of maternal hypothyroidism on embryogenesis, placental development, and the course of the pregnancy. Recent data of Gardner et al. (1978) showing that the uterus in estrogen-treated hypothyroid young rats develops less than in euthyroid animals lend support to this hypothesis.

Comparison of thyroid status in the mother and the newborn: Another original contribution of this study was the demonstration that the degree of abnormality in thyroid function in the newborn is directly limited by the thyroid status in the mother. This has not been observed in normal pregnancy. In fact, the absence of any correlation between parameters of thyroid function in the mother and child led Fisher et al. (1969, 1970) to propose that the pituitary–thyroid regulatory mechanisms were autonomous. A possible explanation for the correlations observed in Ubangi is transplacental passage of maternal T_4 and TSH, but this interpretation is unlikely in view of the clinical and experimental data showing transfer to be negligible (reviewed by Fisher et al. 1977). The most probable explanation is that in severe endemic goitre the mother and fetus are independently and simultaneously exposed to environmental factors that inhibit thyroid function, i.e., deficient iodine supply and elevated serum thiocyanate levels. However, one cannot exclude the possibility that the hypothalamus–pituitary–thyroid axis of the mother and fetus can interact through transplacental transfer, with thyrotropin-releasing hormone (TRH) acting as mediator (Azukizawa et al. 1976). The observation that more newborns than mothers exhibited abnormal TSH values at delivery indicated a more effective regulatory mechanism in the mother than in the newborn. It also indicated that in regions of severe endemic goitre, the perturbations of thyroid function in the mother lead to marked acceleration in thyroid iodine uptake. Thus a direct competition could be established between maternal and fetal thyroid glands. In animals, thiocyanate retards the passage of iodine across the placenta (London et al. 1964), and the same phenomenon probably operates in humans. Consequently, the fetal thyroid gland has great difficulty in securing sufficient iodine.

Delay in psychomotor development: A crucial observation in this work was the demonstration of a diminished development quotient in Ubangi infants of untreated mothers. This difference was observed in the 4–9-month and 10–15-month age groups. A significant difference was also observed in the frequency of success in performing a series of tasks or specific items in the Brunet-Lézine scale. These differences were noted in two groups of children who were examined double-blind and who were entirely comparable with respect to number of siblings, age of mother, etc. The only difference between the two groups was that iodized oil had been administered to half the mothers. Thus, our data confirmed in a particularly rigorous manner, the preliminary observations from Peru (Pretell et al. 1972) and Ecuador (Dodge 1969; Fierro-Benitez et al. 1974a) on mental retardation in severe endemic goitre.

The absence of a significant difference in psychomotor development between 16 and 23 months has been difficult to interpret. It could be due to the limited number of children studied or because the test loses some of its validity at this age. The latter hypothesis gains support from the frequency with which this age group refused to perform the tasks. In the youngest group of both treated and untreated subjects, the psychomotor quotient was elevated — a finding that is consistent with data in African children reported by others (Senecal and Falade 1956; Geber et al. 1957a, b; Dasen 1974).

We have concluded from this study that congenital hypothyroidism and global decrease in thyroid function in the newborn in Ubangi are associated not only with increased incidence of cretinism but also with retardation in psychomotor development in all the children. This conclusion is particularly noteworthy in that abnormalities discovered at this age have an important prognostic significance for later mental development (Lubchenco et al. 1963; Drillien 1972).

Chapter 10

Effects of Thiocyanate During Pregnancy and Lactation on Thyroid Function in Infants

F. DELANGE, P. BOURDOUX, R. LAGASSE, A. HANSON, M. MAFUTA, P. COURTOIS, P. SEGHERS, AND C.H. THILLY

The question may be asked whether SCN influences the thyroid function of the fetus, newborn, and infant, independent of direct ingestion of cassava. Indeed, it has been shown in animals that SCN crosses the placenta (Boulos et al. 1973), inhibits active concentration of iodide in the placenta (Logothetopoulos and Scott 1955) and the mammary glands (Brown-Grant 1956, 1957, 1961; Potter et al. 1959; Gross 1962; Piironen and Virtanen 1963; Wolff 1964; Wood 1975), and exerts a goitrogenic effect in newborns and lactating pups. This last point is discussed in detail in chapter 11.

In contrast, there is little information concerning the metabolism of SCN during pregnancy and lactation in humans (Funderburk 1966; Funderburk and van Middlesworth 1967; Andrews 1973; Pettigrew et al. 1977).

The object of our work was to evaluate whether, in Ubangi, the newborn and infant are exposed to SCN overload of maternal origin and whether this overload influences thyroid function in the newborn.

We measured serum concentration of SCN, TSH, and thyroid hormones in five groups of subjects in Ubangi: women during pregnancy and at delivery, newborns, lactating mothers, and breast-fed infants. The study was completed by measuring SCN in mother's milk. The results were compared with those found in similar groups in Belgium used as controls. In addition, we studied the relations between the maternal urinary iodide/thiocyanate ratio during pregnancy and the serum levels of TSH, T_4, and T_3 in the newborn.

Subjects and Methods

The study comprised 840 subjects — 619 from Zaire and 221 Belgian controls (tables 53–55).

The majority of the Zairian pregnant women, newborns, and infants participated in the study of thyroid function in pregnancy and fetomaternal relations described in chapter 9. The pregnant women did not receive iodized oil during pregnancy. The lactating mothers were investigated in 15 different villages situated in the region of Gemena. Mother's milk was collected from 2 weeks to 2 months after delivery.

The normal Belgian subjects were investigated in the obstetrical and pediatric departments of Hôpital Saint-Pierre in Brussels. Mother's milk was collected on the 5th day after delivery. The infants investigated in Zaire were aged 2 weeks to 15 months; those in Belgium 1–30 months. The mean age of Zairian infants was 6.4 months, of Belgian infants 10.4 months ($P > 0.5$).

The methods for measuring serum SCN, TSH, T_4, and T_3; urinary iodide; and SCN have been detailed in chapter 5. SCN in mother's milk was determined as in serum after precipitation of milk proteins with trichloroacetic

Table 53. Comparison of the serum levels of thiocyanate (SCN), thyrotropin (TSH), thyroxine (T_4), and triiodothyronine (T_3) in Zairian and Belgian (control) mothers at delivery, newborns, and infants.[a]

Serum levels	Pregnant women in Zaire	Mothers at delivery		Newborns		Infants	
		Belgium	Zaire	Belgium	Zaire	Belgium	Zaire
SCN (mg/dl) (mean ± SE)	0.80 ± 0.04 (134)	0.21 ± 0.02 (47)	0.68 ± 0.04*** (68)	0.19 ± 0.02 (47)	0.62 ± 0.04*** (53)	0.38 ± 0.03 (54)	0.55 ± 0.04*** (54)
TSH (μU/ml) (mean ± SE)	1.53 ± 2.1 (221)	3.9 ± 0.4 (47)	14.0 ± 2.1*** (86)	6.8 ± 0.9 (40)	50.9 ± 7.6*** (85)	3.5 ± 0.2 (56)	52.6 ± 11.6*** (56)
T_4 (μg/dl) (mean ± SE)	8.0 ± 0.4 (202)	15.3 ± 0.7 (44)	8.6 ± 0.6*** (76)	12.0 ± 0.5 (38)	7.5 ± 0.4*** (76)	8.7 ± 0.3 (53)	6.9 ± 0.7* (54)
T_3 (ng/dl) (mean ± SE)	206 ± 4 (207)	220 ± 8 (45)	191 ± 9* (81)	60 ± 20 (39)	106 ± 91*** (84)	181 ± 9 (55)	185 ± 7 NS (56)

[a]Figures in parentheses are the number of subjects. Significance: NS = not significant; * = P <0.05; *** = P <0.001.

acid 25% volume/volume and chloroform extraction of supernatant lipids.

Results

The results obtained in Zairian and Belgian subjects of measurements of serum SCN, TSH, T_4, and T_3 in women during pregnancy and at delivery, newborns and infants are compared in Table 53. The mean SCN levels in the four Zairian groups were 1.5–3 times those in the Belgian controls ($P<0.001$). This modification was associated with a very significant elevation in serum TSH and decrease in T_4 ($P < 0.001$) except in infants where the difference in T_4 values between Zairians and Belgians was at the limit of statistical significance ($P<0.05$). In contrast, the serum T_3 concentrations were not significantly modified in the Zairian subjects ($P > 0.5$) except in women at delivery where they were slightly lowered ($P < 0.05$) and in newborns where they were increased 1.5-fold in comparison with Belgian controls ($P < 0.001$).

The individual values for serum SCN in women at delivery and in their newborns are compared in Fig. 68. Data from both Ubangi and Belgium are presented. The values in Belgian subjects were clearly lower than in Zairians. However, some Belgian mothers were heavy smokers, and their newborns demonstrated serum SCN levels equal to those in Zairians. In the Zairian subjects there was a large dispersion in individual values from 0.16 to 1.33 mg/dl. For all the Zairians and Belgians, there was a highly significant correlation between the levels in the mother at delivery and in the cord blood of the newborns ($r = 0.930$, $P < 0.001$).

Table 54. Comparison of urinary iodide and thiocyanate concentrations in Zairian women during pregnancy and at delivery.[a]

Urinary concentration	Pregnant women	Women at delivery
^{127}I (μg/dl) (mean ± SE)	3.4 ± 0.2 (185)	2.4 ± 0.2*** (80)
SCN (mg/dl) (mean ± SE)	2.4 ± 0.1 (126)	1.5 ± 0.2*** (56)
SCN/I ratio (mg/μg) (mean ± SE)	0.90 ± 0.07 (122)	0.86 ± 0.11 NS (37)

[a]Figures in parentheses are number of subjects. Significance: NS = not significant; *** = P <0.001.

The results obtained for the urinary concentrations of iodide and SCN and for the SCN/I ratio in Zairian women during pregnancy and at delivery are compared in Table 54. The iodide and SCN concentrations were lower at delivery than during pregnancy (P<

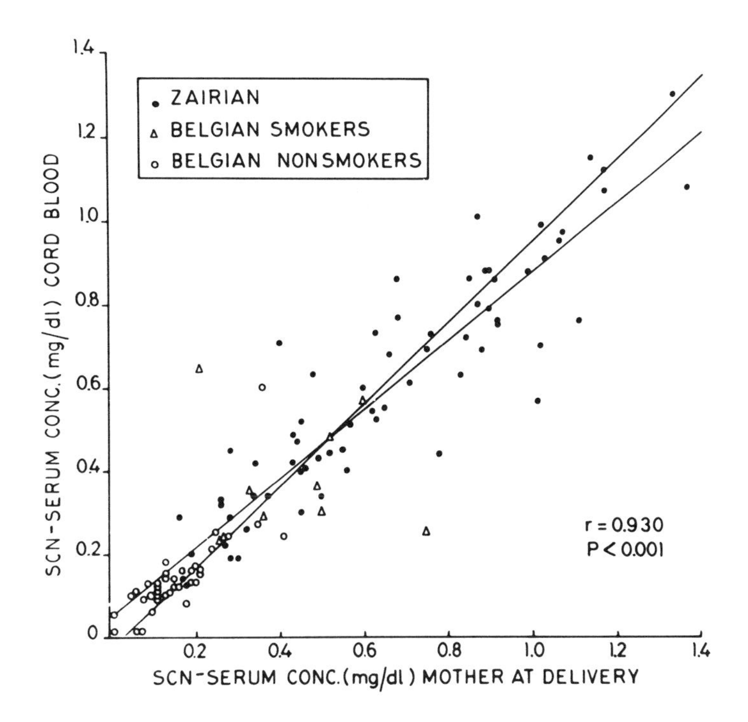

Fig. 68. Relationship between the serum levels of thiocyanate in women at delivery and in their newborns.

0.001). In contrast, the SCN/I ratio was not significantly different ($P > 0.5$). The values of 0.90 and 0.86 were of the same order of magnitude as those presented in chapters 4 and 6 for the general population of this region.

There was a direct correlation between the urinary I/SCN ratio in the mother during pregnancy and the T_4 concentration in cord blood ($r = 0.595$, $P < 0.001$) and an inverse correlation between this ratio and the cord blood TSH concentration ($r = -0.264$, $P < 0.05$). However these correlations were not more significant than those observed between the urinary iodide concentration in the mother and the levels of TSH and T_4 in the newborns.

There were no important modifications in serum SCN, TSH, T_4, or T_3 as a function of age in Zairian and Belgian infants ($r = 0.130$, $P < 0.1$), although the T_4 and T_3 levels in Belgian infants decreased slightly ($r = -0.417$ for T_4, $r = -0.360$ for T_3; $P < 0.001$).

The individual values for TSH and T_4 in Zairian and Belgian infants are compared in Fig. 69. In the Zairian infants, there was important dispersion in the values. The TSH and T_4 levels were within the range of values in

Table 55. Comparison of the thiocyanate (SCN) concentrations in serum and in human milk and of the milk/serum SCN ratio in Zairian and Belgian (control) lactating mothers.[a]

Concentration	Belgians	Zairians
Serum SCN (mg/dl) (mean ± SE)	0.23 ± 0.03 (18)	1.10 ± 0.05*** (140)
Human milk SCN (mg/dl) (mean ± SE)	0.26 ± 0.02 (71)	0.33 ± 0.02 NS (171)
Milk/serum SCN ratio (mean ± SE)	1.26 ± 0.13 (18)	0.48 ± 0.09** (139)

[a]Figures in parentheses are number of subjects. Significance: NS = not significant ($P > 0.05$); ** = $P < 0.01$; *** = $P < 0.001$.

control subjects in only 33 of the 56 subjects investigated. In 12 subjects, TSH was higher than 100 μU/ml and T_4 lower than 2 μg/dl. Five of these 12 infants presented clinical signs of thyroid insufficiency.

The SCN concentrations in serum and mother's milk and the ratio between these concentrations in Zairian and Belgian lactating women are compared in Table 55. Despite the fourfold elevation in serum SCN in Zairian women ($P < 0.001$), the SCN content in milk was barely increased as compared with Belgian women ($P > 0.05$). The ratio of the SCN concentration in milk and serum was 0.48 in Zairian women and 1.26 in Belgians ($P < 0.01$).

Discussion

This work showed that in the Ubangi area:

- The serum thiocyanate concentrations are extremely elevated in pregnant and lactating women; thiocyanate is elevated as well in newborns and breast-fed infants before cassava is introduced into the diet in important amounts;
- This modification is associated with decreased serum T_4 and increased TSH, both in newborns and breast-fed infants; the levels attained are characteristic of severe thyroid insufficiency in approximately one of four infants; nearly half of these infants present typical clinical and radiologic signs of myxedematous cretinism (Delange et al. 1976); and
- The degree of alteration of thyroid function in the newborn is directly related to the balance of dietary supply of iodide and thiocyanate in the mother during pregnancy.

The increased serum SCN level observed in cord blood resulted from transplacental diffusion of the ion. The SCN concentration in the newborn was directly proportional to that in the mother, with a range from normal values of the order of 0.1 mg/dl to extremely high levels surpassing 1.0 mg/dl. This observation confirmed analogous results reported by Andrews (1973) in occidental newborns whose mothers had elevated serum SCN concentrations because of heavy smoking.

The elevated serum SCN in Zairian infants is not explained by the SCN concentration in mother's milk. Our work confirmed that in women (Funderburk and van Middlesworth 1967), as in cows (Virtanen and Gmelin 1960; Virtanen 1961), the SCN content of mother's milk is not proportional to its concentration in the serum of lactating mothers as it is in female rats at the end of lactation (Funderburk and van Middlesworth 1967). However, it is interesting

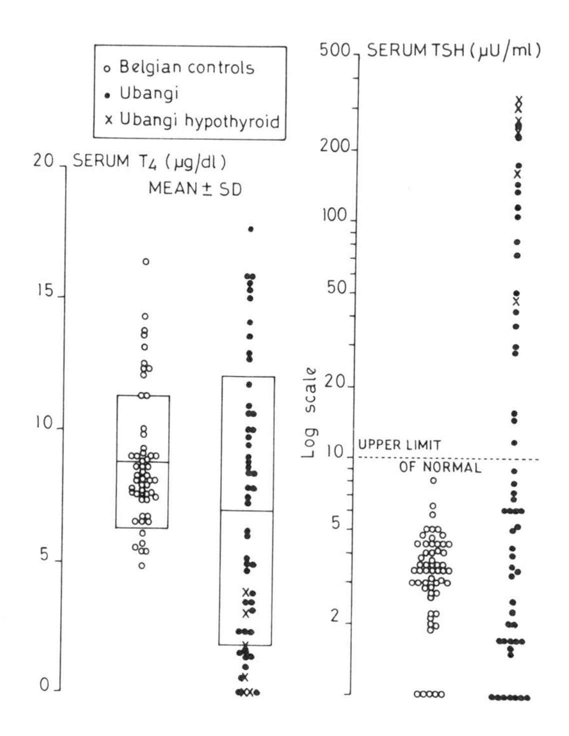

Fig. 69. Comparison of the serum levels of T_4 and TSH in Zairian (Ubangi area) and Belgian infants.

to note that the amount of SCN in mother's milk in Zaire was not negligible. On the basis of a mean SCN concentration of 0.3 mg/dl (Table 55) and a daily intake of approximately 800 ml milk, the breast-fed infant receives 2–3 mg thiocyanate/day. In an infant 7–9 kg, with a SCN distribution space of approximately 30%

of the body weight (Eder 1951), this amount may cause an increase in serum SCN of 0.1 mg/dl/day. This mechanism probably explains the higher serum SCN concentrations in infants than in newborns or adults.

The high serum SCN in Zairian infants could also be explained by a persistence of SCN crossing the placenta. However, this hypothesis seems unlikely as the half-life of SCN in humans is 1–14 days (Eder 1951; Pettigrew and Fell 1972; Butts et al. 1974; Bourdoux et al. 1978; Schulz et al. 1979). A more plausible explanation is that breast-fed infants in this area are given a supplement of cassava juice during lactation (unpublished data).

The elevation in serum SCN concentration observed in the pregnant and lactating mother and in the newborn and infant in Ubangi may interfere with thyroid function of the fetus and young infant in two ways:

- Thiocyanate ion may act directly on the thyroid by a mechanism identical to that in the adult.
- The ion may also inhibit active iodide transport in the placenta and mammary glands, i.e., in the two organs that determine the iodine supply in the fetus and breast-fed infant (Weaver et al. 1960; Macy and Kelly 1961; Ling et al. 1961, Knowles 1965; Man and Benotti 1969).

Under physiologic conditions, the iodide concentration in mother's milk is 6–20-fold higher than in serum (Honour et al. 1952; Funderburk and van Middlesworth 1967).

The result of these inhibitory mechanisms is a reduced availability of maternal iodide under conditions in which the mothers are already subjected to the iodine deficiency prevalent in the population. This situation exists precisely during the critical period of growth in the fetus and infant when the thyroid constitutes its iodine stores (Etling 1977; Ketelbant-Balasse et al. 1978), explaining in part the very low thyroid iodine content in endemic goitre, particularly in the young infant (Delange 1974).

Under these precarious conditions, the gland is probably at the limit of its functional capacities. Any additional hindrance of optimal intrathyroidal metabolism may lead to decompensation of the adaptive mechanisms and result in thyroid insufficiency. Such a situation has recently been demonstrated in sick premature infants in Belgium in whom iodine supply is still higher than that found in Zairian infants (Delange et al. 1978). The hypothesis is further supported by the observation that thyroid function is normal in newborns of mothers receiving iodine supplement during pregnancy (Delange et al. 1976; Thilly et al. 1978; chapter 9).

Our data were not sufficient to establish a decisive role for SCN in the development of abnormalities in thyroid function in newborns and infants. The abnormalities were certainly due in large part to iodine deficiency. The demonstration is difficult because of the interdependence of the two mechanisms. As shown in chapters 4, 5, 6, and 11, the role of SCN could only be established in adults because of highly contrasting epidemiological situations and in the rat by extreme experimental conditions. However, the demonstration of elevated serum SCN concentrations in newborns and infants, approaching the values observed in adults, strongly favours the intervention of this goitrogenic factor.

The absence of aggravation of thyroid failure in infants as compared with newborns contradicts the experimental data presented in chapter 11, which showed an important increase in the goitrogenic effect at the end of lactation in the rat. The discrepancy may be due to species differences or perhaps to an increased mortality in hypothyroid infants as suggested by the results in chapter 9. Another possible explanation is that the infant receives thyroid hormones in mother's milk (Brown-Grant and Galton 1958; Strbak et al. 1976, 1978; Sack et al. 1977; Bode et al. 1978; Varma et al. 1978).

In conclusion, this work showed that cassava ingestion by pregnant and lactating mothers results in SCN overload in the fetus, newborn, and infant, even before they begin systematic ingestion of cassava. In analogy to the results discussed in chapters 4, 5, 6, and 11, it is highly probable that this exposure to SCN triggers the development of thyroid insufficiency during the most critical period of adaptation to iodine deficiency. This period corresponds chronologically to the most active phase of differentiation of the central nervous system. Therefore, a logical postulate is that thiocyanate overload in newborns intervenes in the pathogenesis of endemic cretinism.

Chapter 11

Influence of Goitrogens in Pregnant and Lactating Rats on Thyroid Function in the Pups

F. DELANGE, N. VAN MINH, L. VANDERLINDEN, K.D. DÖHLER, R.D. HESCH, P.A. BASTENIE, AND A.M. ERMANS

The action of goitrogenic substances on thyroid function in the adult animal has been extensively studied (reviewed by Studer and Greer 1968; Yamada et al. 1974; Green 1978). In contrast, few studies exist on goitrogenic effects during the fetal and perinatal period (Schmidt and Allen 1967; Davenport 1970; Ekpechi and van Middlesworth 1973; Städtler and Klemm 1973; Greer et al. 1975; Spindel et al. 1976; Kreutler et al. 1978). This period is particularly critical, as the thyroid grows most rapidly and accumulates its iodine reserves during this time (Stolc et al. 1973; Vigouroux 1976; Pascual-Leone et al. 1978). It has been hypothesized that in severe endemic goitre the thyroid is damaged during fetal life or soon after birth in humans and that this damage is responsible for the development of endemic cretinism (Dumont et al. 1963b; Delange et al. 1972c).

The goal of the present work was to evaluate the effect of goitrogenic substances administered to rats during gestation and lactation on thyroid function in their pups from birth to weaning. The experimental model was designed to simulate the situation of human newborns in a zone of endemic goitre. During fetal life and the postnatal period, the infant receives iodine, and eventually goitrogenic substances, by transplacental passage and subsequently from mother's milk.

We have paid particular attention to the possibility of damaging the fetal thyroid to reproduce the lesions suspected of inducing endemic cretinism. Two hypotheses have been tested:

- That exhaustion atrophy secondary to intense stimulation by endogenous fetal TSH causes the damage (Dumont et al. 1963b); and
- That the acute action of an iodine overload in the hyperstimulated gland is responsible. Such a treatment can produce necrosis of thyroid follicles in the dog (Belshaw and Becker 1973).

Our interest in this study was centred on the fetal thyroid. Thus, because of the risk of interrupting the pregnancy as a result of goitrogen-induced hypothyroidism in the mother (Schultze and Noonan 1970), we gave T_4 supplements to the pregnant rats. It has been shown that the placental barrier in the rat is practically impermeable to T_4 and T_3 (reviewed by Fisher and Dussault 1974; Fisher 1975; Fisher et al. 1977), whereas antithyroid drugs and iodide freely pass through it (Nataf et al. 1956; D'Angelo and Wall 1971, Brownlie et al. 1971; Quinones et al. 1972; Städtler and Klemm 1973; Slanina et al. 1973; Book et al. 1974; Kreutler et al. 1978).

Material and Methods

Wistar rats, 2 days pregnant (University of Louvain, Belgium), weighing 175–250 g were divided into four groups, each receiving a dif-

Table 56. Diets given to pregnant rats in an effort to reproduce human situation in Zaire.

Group	Diet
Controls (A)	Low iodine diet (LID) + 5 μg KI/day + 2.5 μg T_4/100 g weight/day (T_4)
Antithyroids (B)	LID + 10 mg SCN/day + 10 mg PTU/day + T_4
Antithyroids + 25 μg KI (C)	LID + 10 mg SCN/day + 10 mg PTU/day + T_4 + 25 μg KI (intraperitoneally)
Antithyroids + 200 μg KI (D)	LID + 10 mg SCN/day + 10 mg PTU/day + T_4 + 200 μg KI (intraperitoneally)

ferent diet (Table 56). The basic regimen was the iodine-poor Remington diet. In the control group (A), the iodine deficiency was corrected by the administration of 5 μg KI/day. In the other groups, the effects of iodine deficiency were enhanced by the administration of 10 mg thiocyanate (SCN) and 10 mg propylthiouracil (PTU)/day. To evaluate the immediate effect of an iodine supplement on fetal thyroid function, our team injected two groups (C and D) intraperitoneally with a single dose of KI 2 days before the end of gestation, 25 μg in group C, 200 in group D. All the rats received physiological amounts of T_4, 2.5 μg/100 g weight/day (Schultze and Noonan 1970). The KI, SCN, PTU, and T_4 were added to the diet during preparation. Drinking water was provided ad libitum.

The experimental protocol is shown in Fig. 70. The mothers received their diets from the 2nd day of gestation until the end of lactation, i.e., the 16th day of life of the pups. Half of the mothers and offspring were killed immediately after delivery. The other mothers and pups, whose only source of nourishment was mothers' milk, were killed at weaning, i.e., on the 16th day. The mothers were anesthetized with ether. The pups were sacrificed by rapid cooling in ice. Blood was collected by intracardiac puncture. The thyroids were dissected, immediately weighed and preserved in Bouin's solution.

The investigation comprised 37 mother rats and 257 pups.

Serum levels of TSH were determined by means of a NIAMDD radioimmunoassay kit. Rat TSH-1-2 as ^{125}I-labeled antigen, anti-rat TSH-S-3 as antibody and rat TSH-RP-1 as reference standard were used for TSH radioimmunoassay procedures. Serum levels of T_3 and T_4 were measured according to a modified version of the method described by Mitsuma et al. (1972). Antibodies for T_3 and T_4 were raised by Hehrmann (Hehrmann and Schneider 1974) according to a modified version of the method described by Hesch and Hüfner (1972). Free and bound antigens were separated by a double-antibody-solid-phase method (DASP from Organon) as described by Hesch et al. (1974).

The sections were stained with PAS-hematoxylin. The morphological findings were classified into four categories (Table 57), illustrated in Fig. 71.

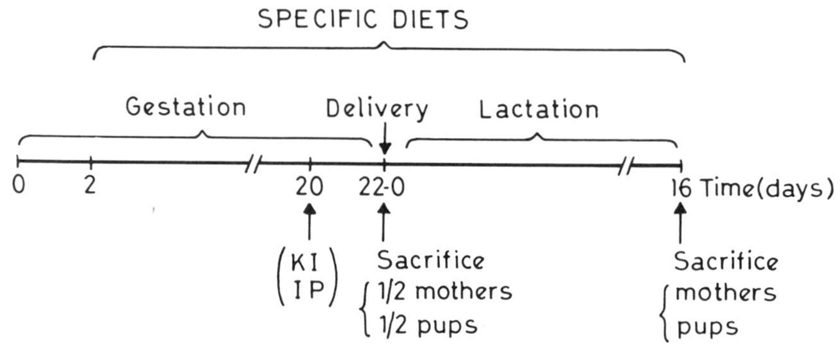

Fig. 70. Experimental protocol.

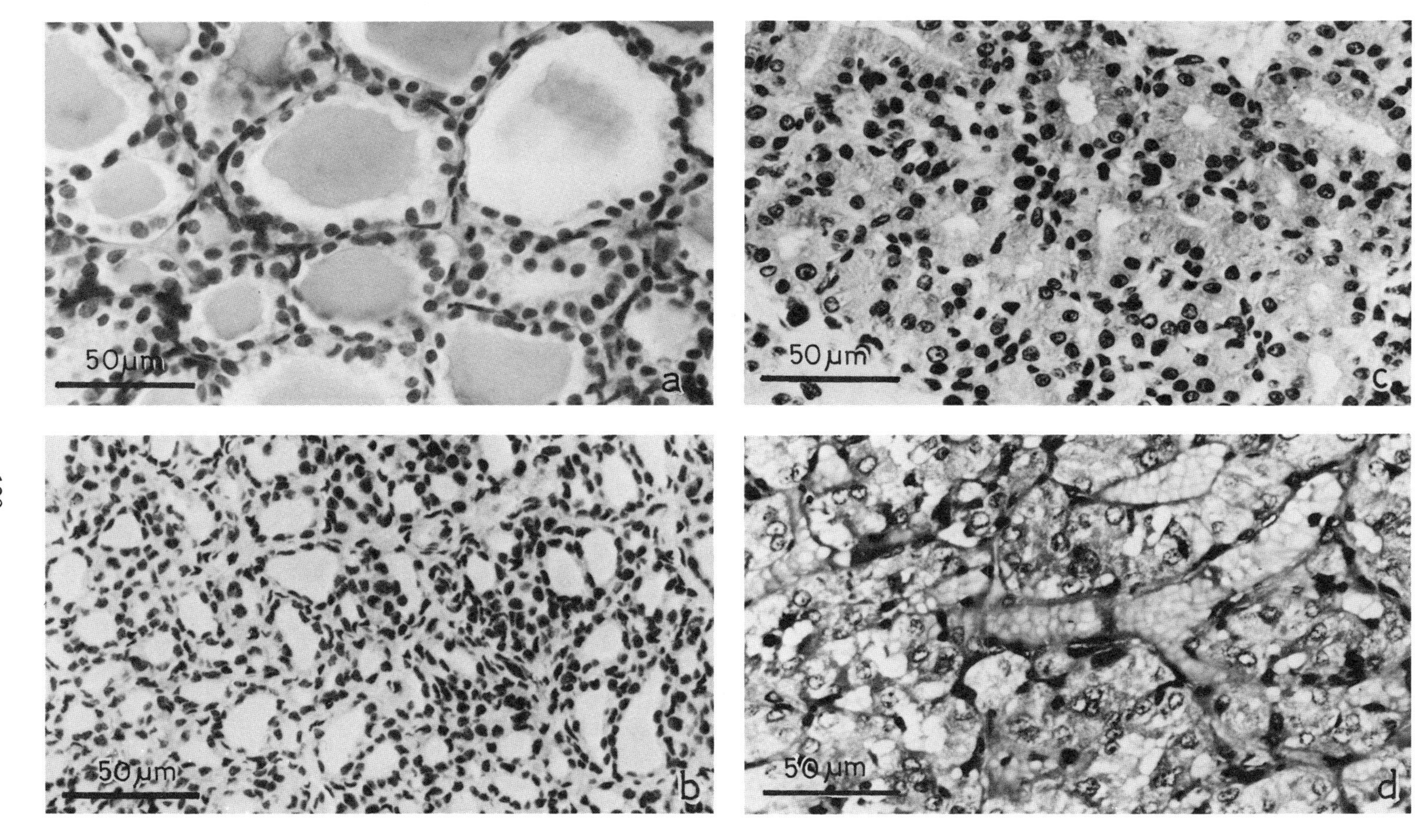

Fig. 71. *Histological characteristics of the thyroid gland: (a) normal in mother from the control group (A) at delivery; (b) slightly stimulated (I) in pup from controls (A) at 16 days; (c) stimulated (II) in pup from antithyroid group (B) at delivery; and (d) hyperstimulated (III) in mother from antithyroid group + 25 µg KI at 16 days.*

Table 57. Histological characteristics of thyroid gland at different levels of stimulation.

Category	Level	Vesicles	Epithelium	Colloid	Mitosis
Normal	0	Large	Flat	Abundant	0
Slightly stimulated	I	More or less large	Cubic	Reduced	0 or +
Stimulated	II	Small	Cylindric	Absent	+
Hyperstimulated	III	Very small or no visible vesicular structure; compact aspect	High cylindric, with or without extravesicular hyperplasia	Absent	+++ or 0 (postmitotic hyperplasia)

Results

The weights of the mother rats and pups at delivery and weaning are shown in Table 58. The results in the experimental groups (B, C, D) were not significantly different from those in the control group (A) ($P > 0.5$) with the exception of the group C pups at weaning. Their mean weight was significantly lower ($P < 0.01$).

The weights of the thyroids in the four experimental groups are compared in Fig. 72.

Mothers: In spite of supplementary T_4, the administration of antithyroid drugs (group B) induced considerable thyroid hyperplasia as compared with the control group, both at the time of delivery ($P < 0.001$) and at weaning ($P < 0.01$). The hyperplasia was less evident after administration of 25 μg KI (group C) and 200 μg KI (group D). In the four groups, the thyroid weights at birth and at 16 days were not significantly different ($P > 0.1$–$P > 0.4$).

Pups: The results at birth were similar to those observed in the mothers. However, in the offspring, the thyroid hyperplasia was less marked.

In contrast, at 16 days, the thyroid weights in the pups were 3–4 times those at birth. The difference in weight at these two times was highly significant ($P < 0.001$). In addition, the thyroid hyperplasia observed at weaning was not corrected by prior administration of 25, or even 200, μg KI to the mothers.

The results of histological examination of the thyroids are summarized in Table 59. In the control group (A), the thyroids of the mothers and pups did not appear stimulated, although in a few pups slight hyperactivity was observed. The considerable hyperplasia observed in group B, secondary to the administration of antithyroid drugs was accompanied by microscopic changes of stimulation or hyperstimulation in all animals. These morphological changes were not reversed by the administration of 25 μg KI (C). However, after 200 μg KI

Table 58. Comparison of weights of mothers and pups at delivery and at weaning 16 days later in the four groups of rats.[a]

Group	Weight of mothers (g) (mean ± SE) at: Delivery	Weight of mothers (g) (mean ± SE) at: Weaning	Weight of pups (g) (mean ± SE) Delivery	Weight of pups (g) (mean ± SE) 16 days
Controls (A)	196.0 ± 8.5 (7)	205.3 ± 13.0 (4)	4.8 ± 0.1 (52)	17.7 ± 0.6 (25)
Antithyroids + T_4 (B)	195.1 ± 5.8 NS (5)	221.8 ± 14.7 NS (5)	5.1 ± 0.1 NS (36)	16.5 ± 0.5 NS (25)
Antithyroids + 25 μg KI + T_4 (C)	177.6 ± 7.8 NS (3)	200.8 ± 8.1 NS (4)	5.1 ± 0.5 NS (24)	12.3 ± 0.4*** (35)
Antithyroids + 200 μg KI + T_4 (D)	191.8 ± 2.9 NS (4)	207.9 ± 6.5 NS (5)	5.3 ± 0.1 NS (34)	16.2 ± 0.5 NS (26)

[a]Figures in parentheses are numbers of animals. Significance: NS = not significant; *** = $P < 0.001$.

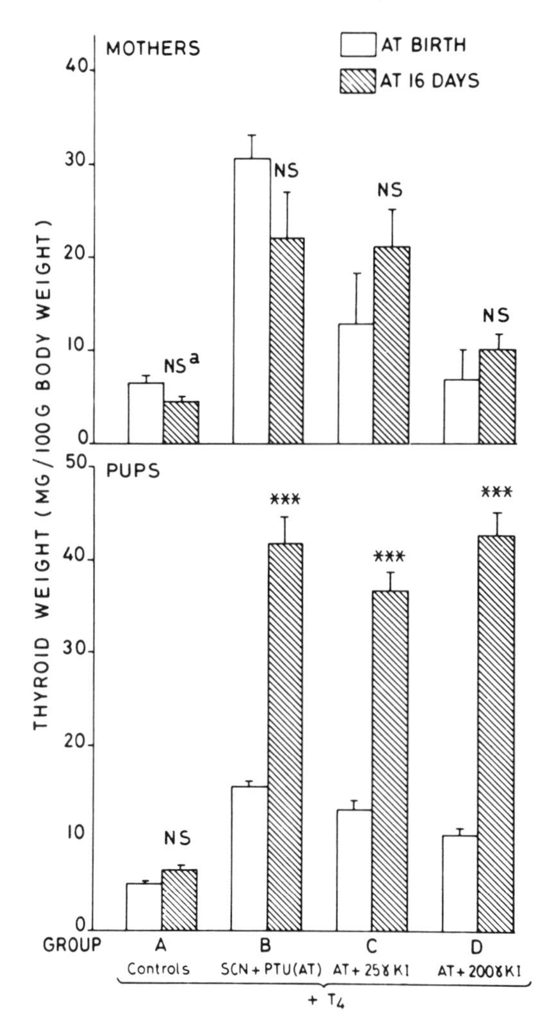

*Fig. 72. Comparison of the weights of the thyroids at birth and at 16 days in the mothers and in the pups in the four experimental groups. Levels of significance as compared with the results obtained at birth: NS = Not significant; *** = $P < 0.001$.*

(D) the thyroid glands appeared to be in the basal state in all animals except the pups at 16 days where hyperstimulation persisted.

The serum levels of thyroid hormones and of TSH are presented in Table 60.

Mothers: In group B, the levels of T_4 and T_3 at delivery and at weaning were lower than in the control group. However, the differences were significant only for T_4 at delivery ($P < 0.001$) and for T_3 at weaning ($P < 0.05$). Administration of 25 or 200 μg KI (C, D) led to normalization of serum thyroid hormone concentrations ($P > 0.5$). At weaning, serum TSH levels were increased in groups C and D, but the differences were not statistically significant ($P > 0.1$).

Pups: In group B at birth T_4 levels were decreased as compared with the controls ($P < 0.001$). However in contrast to the results obtained in the mothers, the serum T_4 level was not corrected by KI administration (C, D), and serum TSH was increased approximately 10-fold over values in the control group ($P < 0.001$). The T_3 levels did not differ significantly in the four experimental groups ($P > 0.5$).

Discussion

In this experimental model, the administration of T_4 to the pregnant rats enabled them to produce litters in which the numbers and weights of the pups were practically normal despite the large doses of antithyroid drugs administered to the mothers. The serum T_3 levels in the mothers were maintained — a finding that is generally accepted as indicative of the euthyroid state (Studer and Greer 1968). The T_4 supplement corresponded to physiologic needs, 2.5 μg/100 g body weight/day (Schultze and Noonan 1970; Gray and Galton 1974). Nevertheless, administration of SCN and PTU to pregnant animals induced considerable thyroid hyperplasia at delivery and at weaning. This observation is in agreement with the results of other authors showing that T_4 does not prevent the goitrogenic effects of certain antithyroid agents except in supraphysiologic doses (van Middlesworth et al. 1959; Jagiello and McKenzie 1960) and that under certain conditions it can even potentiate them (Sellers and Schönbaum 1962). We do not have an explanation for the lowered serum T_4 observed in the mothers at the time of delivery in contrast to the normal level at weaning.

The administration of a single injection of KI to the pregnant rats just before delivery induced nearly complete normalization of thyroid gland weight. This observation supports the hypothesis that the goitrogenesis is

Table 59. Histological aspects of thyroid in mothers and pups at delivery and at weaning as a function of diet during pregnancy.

Group	At delivery				At 16 days (weaning)			
	Mothers		Pups		Mothers		Pups	
	Number of animals	Histological aspect	Number of animals	Histological aspect	Number of animals	Histological aspect	Number of animals	Histological aspect
Controls (A)	7	0	39	0–I	4	0	16	0–I
Antithyroids (B)	4	III	33	II–III	5	III	25	III
Antithyroids + 25 μg KI (C)	3	III	24	III	4	III	35	III
Antithyroids + 200 μg KI (D)	4	I	12	0–I	5	I	21	III

most closely linked to the iodine content of the thyroid (Ermans 1969).

The present study confirms the observations of others who have reported thyroid hyperplasia, at birth, in pups of mothers receiving a goitrogenic diet during pregnancy (Ekpechi and van Middlesworth 1973; Städtler and Klemm 1973; Greer et al. 1975). In our study the hyperplasia, which was explained in part by iodine deficiency and in part by transplacental passage of antithyroid substances became far more marked (3–4 times) in the pups at weaning than at birth. This finding was not reported by other authors investigating the goitrogenic effects of an iodine-deficient diet without antithyroid agents (Ekpechi and van Middlesworth 1973; Greer et al. 1975). The pronounced hyperplasia at 16 days indicates either that during lactation the thyroid of the pups received even less iodide than during fetal life or that the goitrogenic effect of the antithyroid drugs was more pronounced.

Two factors may be operative in the diminution of iodide supply during lactation:

- The paucity of iodide in mother's milk due to the low maternal plasma iodide concentration (Lorscheider and Reineke 1972); and
- A thiocyanate-induced (Brown-Grant 1957; Gross 1962) inhibition of the mammary glands' capacity to concentrate iodide (Potter and Chaikoff 1956; Brown-Grant and Galton 1958; Potter et al. 1959; Grosvenor 1960).

The relative importance of these two factors cannot be determined, and they probably intervene simultaneously.

The hypothesis of a more pronounced effect of antithyroid agents during lactation is less likely. In the rat, the concentration of SCN in the milk is not greater than in serum during most of lactation (Funderburk and van Middlesworth 1967). In addition, we know of no evidence that antithyroid drugs of the PTU type enter mother's milk in the rat.

The failure of the injected iodide to correct the thyroid hyperplasia in the pups at 16 days also suggests that antithyroid agents aggravated the iodine deficiency. In contrast, this treatment induced a marked reduction in the volume of the thyroid in the mothers and pups at delivery. The greater thyroid hyperplasia observed in the pups at weaning, associated with biochemical signs of thyroid insufficiency, likewise favours this hypothesis.

It appears that, under the experimental conditions used, the transfer of iodine in the milk is reduced and, in particular, that the SCN interferes with the mammary iodide pump and that the maternal thyroid exhibits extreme iodide avidity. The result is that the pups develop an exceptionally severe iodine deficiency and that they are incapable of bene-

Table 60. Comparison of serum levels of TSH, T_4, and T_3 in rats at delivery and at weaning (16 days).[a]

Group	Mothers at delivery		At 16 days: Mothers			At 16 days: Pups		
	T_4 (μg/dl)	T_3 (ng/ml)	T_4 (μg/dl)	T_3 (ng/ml)	TSH (ng/ml)	T_4 (μg/dl)	T_3 (ng/ml)	TSH (ng/ml)
Controls (A)	1.25 ± 0.21 (7)	0.39 ± 0.08 (7)	1.83 ± 0.17 (5)	0.42 ± 0.06 (5)	80 ± 22 (4)	3.49 ± 0.15 (40)	0.28 ± 0.04 (40)	240 ± 18 (6)
Antithyroids (B)	0.49 ± 0.05** (5)	0.22 ± 0.04 NS (5)	1.41 ± 0.14 NS (5)	0.19 ± 0.04* (5)	-	0.43 ± 0.01*** (34)	0.34 ± 0.02 NS (29)	-
Antithyroids + 25 μg KI (C)	1.59 (2)	-	2.05 ± 0.70 NS (4)	0.44 ± 0.08 NS (4)	783 ± 385 NS (4)	0.44 ± 0.03*** (9)	0.16 ± 0.03 NS (9)	2380 ± 80*** (5)
Antithyroids + 200 μg KI (D)	1.29 ± 0.26 NS (4)	0.39 ± 0.02 NS (4)	2.77 ± 0.66 NS (5)	0.39 ± 0.03 NS (5)	202 ± 65 NS (5)	0.52 ± 0.07*** (12)	0.15 ± 0.01 NS (6)	2320 ± 31*** (8)

[a]Figures in parentheses are numbers of animals. Significance: NS = not significant; * = $P<0.05$; ** = $P<0.01$; *** = $P<0.001$; measurements are expressed as mean ±SE.

fiting from the iodide supplement administered to the mothers.

One cannot exclude the possibility that the absence of a response in the pups is due to a thyroid blockage induced by iodide (Wolff-Chaikoff effect reviewed by Nagataki 1974) at the supraphysiologic dose of 200 μg. However, such an effect is unlikely at the dose of 25 μg, which is closer to the physiologic needs (Inoue and Taurog 1968). Moreover, a Wolff-Chaikoff effect would be expected to appear at birth as well as at 16 days, and it did not.

The histological observations in the pups suggest that in the rat, none of the variables — extreme iodine deficiency, direct action of SCN or PTU, extreme TSH hyperstimulation, and acute iodine overload in the hyperstimulated gland — are capable of inducing thyroid lesions detectable during fetal life or during the first days of life. Extrapolating these conclusions to humans is hazardous because the timing of organogenesis is different in the two species (Stolc et al. 1973; Fisher et al. 1977).

In conclusion, this study showed that the administration of SCN and PTU to pregnant rats receiving an iodine-deficient diet has a goitrogenic effect on the pups, detectable at birth but far more impressive at weaning. At 16 days the thyroid hyperplasia is associated with morphological changes of hyperstimulation and biochemical signs of thyroid insufficiency. This situation seems to be the result partly of the iodine deficiency and partly of the inhibitory action of the antithyroid drugs on the iodide concentration mechanism of the maternal mammary gland. These data suggest that, in humans also, the fetal and, particularly, the lactation period are critical for the action of goitrogenic substances on thyroid function in the presence of severe iodine deficiency.

Chapter 12

Continuous Spectrum of Physical and Intellectual Disorders in Severe Endemic Goitre

R. LAGASSE, G. ROGER, F. DELANGE, C.H. THILLY, N. CREMER, P. BOURDOUX, M. DRAMAIX, D. TSHIBANGU, AND A.M. ERMANS

In populations affected by severe endemic goitre, a large number of patients exhibit anomalies of physical and intellectual development. These patients are called endemic cretins (chapter 4; reviewed by Delange et al. 1976).

Moreover, surveys in endemic goitre areas presenting cretinism indicate that many inhabitants, considered clinically as noncretin, exhibit impairment of intellectual and neurologic development and of thyroid function to various degrees (Ramirez et al. 1969; Ibbertson et al. 1971; Green 1973; Fierro-Benitez et al. 1974a, b; Roger et al. 1977). These observations suggest that in such regions there exists a continuous spectrum of anomalies of which cretinism, as usually defined, represents only the extreme form (Delange et al. 1972b).

The purpose of our study was to evaluate this hypothesis. To do so, we compared intellectual performance and biochemical parameters of thyroid function in four groups of subjects living in a zone of endemic goitre of extreme severity in central Africa (Thilly et al. 1977). We grouped the subjects, according to a single clinical evaluation, into:

- Normal subjects;
- Subjects presenting mild clinical signs of thyroid insufficiency or impairment of mental development or both. These subjects were called "cretinoids";
- Typical myxedematous cretins; and
- Neurologic cretins.

Patients and Methods

The present investigation was carried out in Ubangi-Mongala in northwest Zaire (Thilly et al. 1977), a region of extremely severe endemic goitre covering ~ 200 000 km² with a population of approximately 1.5 million. The prevalence of goitre varies between 39 and 68% of the population and of cretinism between 0.5 and 10%. Myxedematous cretinism is the most frequently observed form of cretinism in this region. The ratio between myxedematous and neurologic cretins is around 11.5:1 in a sample of 131 709 inhabitants with mean prevalences of respectively 1.08% and 0.0934%.

The local population produces most of its own food. Iodine intake is about 10–20 μg per day.

One hundred fifty-four patients were studied including clinically euthyroid subjects, with or without goitre, used as controls; cretinoids; and myxedematous and neurologic cretins.

The investigations were carried out in two phases (Table 61). The first (A) involved 112 patients in 12 villages of the subregion of Ubangi, studied during the mass campaign to supplement inhabitants in this zone by means of iodized oil injections. In this group, clinical examination was completed by a determination of radioactive iodine thyroidal uptake and of TSH, T_4, and T_3 levels in serum. The second phase (B) was carried out in 2 of the 12 villages,

Table 61. Classification of patients from two investigations (A and B).

Group	Characteristics	Number of subjects	
		A	B
Normals	No apparent intellectual, neurologic, or thyroidal impairment	74	9
Cretinoids	Suspicion of hypothyroidism or intellectual impairment or both, on clinical basis	8	4
Myxedematous cretins	Mental defect associated with doubtless myxedema and stunted growth	30	22
Neurologic cretins	Mental defect associated with deaf-mutism and/or defects of stance and gait	–	7

situated near the base of operations of the team, thus allowing a more complete study. In the 42 patients involved in this phase the clinical examination was complemented by a determination of the Achilles reflex time, the mean reaction time, and the bone-age maturation. The intellectual development and the audiometric performances of these patients were also investigated. Their age, sex, height, and number of goitres are presented in Table 62.

All the cretins presented mental retardation recognized by their counterparts and confirmed by brief testing. The 22 myxedematous cretins exhibited, in addition, growth retardation and other clinical signs of covert thyroid insufficiency: myxedema, dry skin, lumbar lordosis, hypogonadism, and facial anomalies (defect in the maturation of the naso-orbital bridge, mandibular hypoplasia, and eversion of the lips). Seven also presented an umbilical hernia approximately 10 cm in diameter. Two cretins had a small stage 1 goitre (Perez classification). In five others, the thyroid gland was not palpable. The age-range of these cretins was 8–26 years with a mean of 16.8. Fig. 73 shows a typical myxedematous cretin.

The seven neurologic cretins showed no signs of hypothyroidism clinically; however, their height was significantly less than that of the controls. Their mean age (17.6) was not significantly different. All had a goitre estimated as stage 1b to stage 3. Hyperactive reflexes were demonstrated in all seven, five of them demonstrating an ataxia or spastic gait, disorders of motor coordination, or strabismus. Four subjects exhibited partial hearing loss. Fig. 74 shows a neurologic cretin.

The four cretinoids were 10–19 years of age (mean 15.1). They were all considered to be mentally handicapped. Their height was significantly less than in control subjects.

Table 62. Parameters characterizing the four groups of subjects.

Parameter	Controls	Cretinoids	Myxedematous cretins	Neurologic cretins
Number	9	4	22	7
Age (y) (mean ± SE)	16.6 ± 0.7	15.1 ± 2.2 NS	16.8 ± 1.2 NS	17.6 ± 1.5 NS
Males/females	6/3	2/2	8/14	4/3
Goitres (n)	5	2	2	7
Height (cm) (mean ± SE)	155 ± 3	123 ± 6***	103 ± 2***	130 ± 5***

[a]Significance: NS = not significant; * = P <0.05; ** = P <0.01; *** = P <0.001.

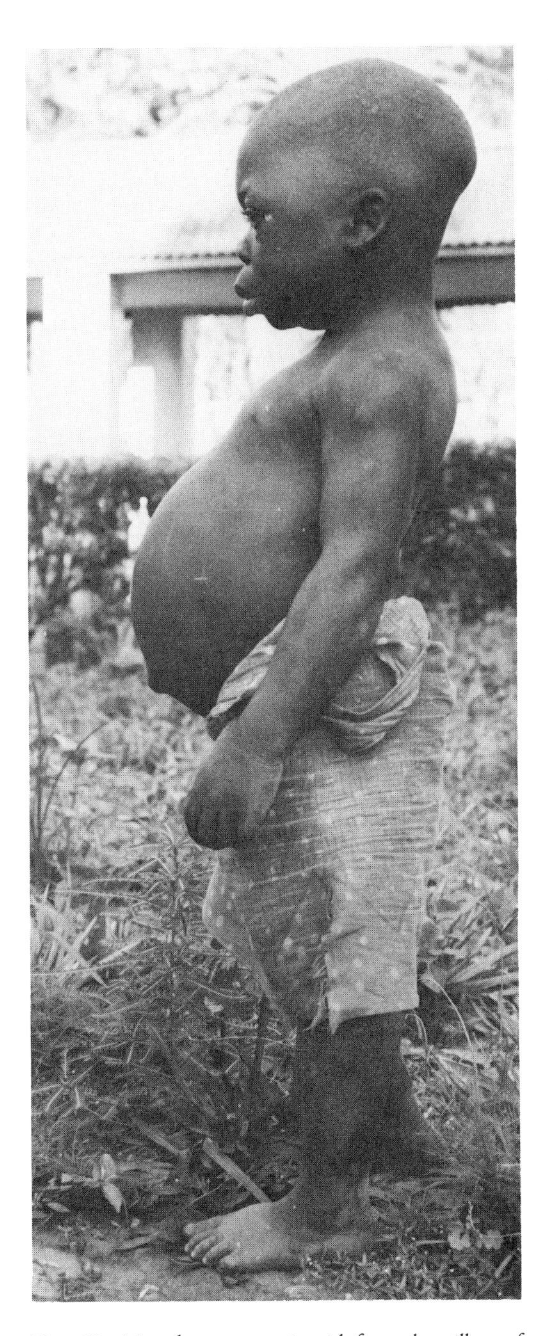

Fig. 73. Myxedematous cretin girl from the village of Bokuda (Ubangi, Zaire); she was 9 years old, 92 cm tall with a bone age of 1½ years. She had no goitre; her T_4 was undetectable; her T_3 was 9 ng/dl; and TSH was 510 μU/ml. The picture shows lumbar lordosis with prominent abdomen and umbilical hernia, as well as the configuration of her face and skull and dryness of skin. She scored 7 a.u. in the Raven test and 12 a.u. in the Gesell test.

The nine control subjects were clinically normal in terms of intellectual, neurologic, and thyroid function. They were chosen from the same age groups as the other categories of patients so that their mean age (16.6) did not differ significantly from the other groups. Five controls had stage 1 goitres.

Parameters of thyroid function: The methods used for measuring serum TSH and T_4 concentrations have been described by Bourdoux et al. (1978). Serum TSH values were assumed to have a log-normal distribution for statistical comparison.

Bone maturation: We determined bone age by examining hand and knee X-ray films. The criteria of Greulich and Pyle (1959) were used for the hand; those of Pyle and Hoerr (1955) for the knee. Retardation of bone maturation was calculated by dividing the difference between the chronological and bone age by the chronological age. The results were expressed as a percentage.

Intellectual development: Intellectual development was evaluated on the basis of two classical tests, slightly modified, and considered exempt from cultural bias:

- A matching test borrowed from Gesell's scale of psychomotor development (Geber and Dean 1957), in which the subject was presented with a tray bearing cutouts of a circle, square, and triangle. The corresponding wooden blocks were offered to the subject in the same order as they appeared on the tray, and he or she was asked to place the blocks in the appropriate holes. The test was repeated after the tray was turned 180° and the order of presentation of the blocks reversed. The original method was modified on two points: to ameliorate the appreciation of the forms, we presented the blocks in an order that did not correspond to the order of the holes; in addition, we introduced a grading system as a function of the frequency of correct responses. The individual scores varied from 1–4 according to the difficulty of the task. The maximum score was 17.

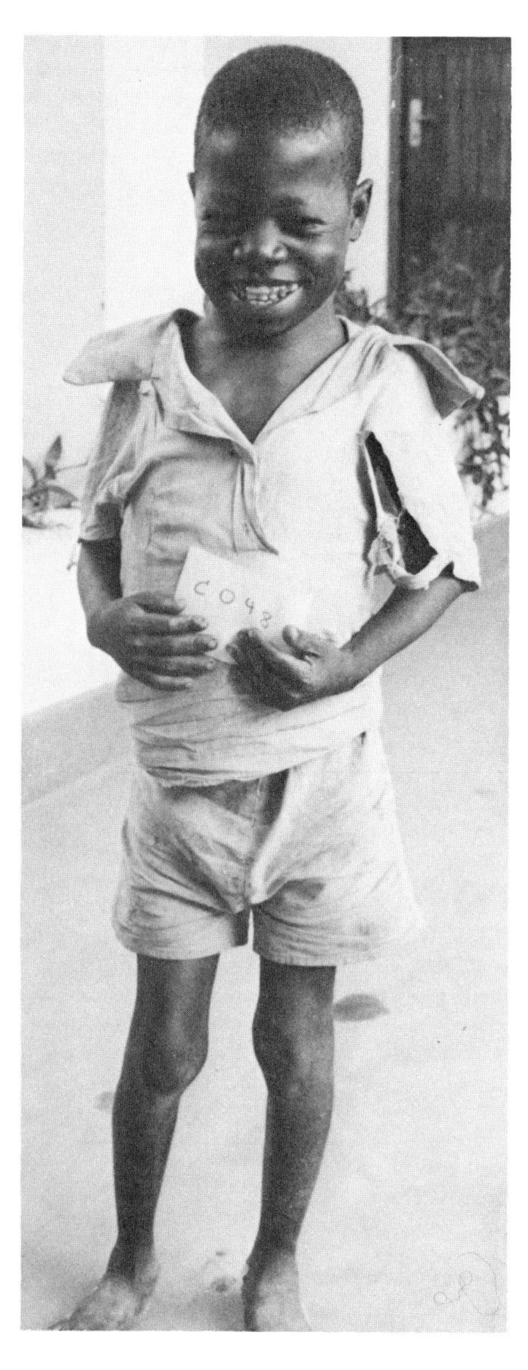

Fig. 74. *Neurologic cretin boy from the village of Bokuda (Ubangi, Zaire) at age 16 years. His height was 121 cm but bone age was not determined. He had stage 1b goitre; T_4, 0.3 μg/dl; T_3, 133 ng/dl; and TSH, 48 μU/ml. This boy was afflicted by a spastic diplegia and extremely impaired mental functions: Raven test, 0 a.u., and Gesell test, 3 a.u.*

• The colour matrix (sets A, A6, P) of J.C. Raven (1965). The goal of this test is to choose from eight possible pieces, the best one to complete a geometric figure. The figure is coloured, and the pieces vary only by their colours and geometric pattern, not by their form as in the preceding test. Testing was based on the method of Ombredane (Ombredane et al. 1956). The subjects were tested individually and were asked to point a finger at the right answer. The test was preceded by a demonstration and a trial test by the subject. The protocols were graded in the classic manner, and the final score was equal to the number of the correct responses, the maximum being 36. This test is more difficult than the first, thus allowing a discrimination of subjects with maximal scores in the Gesell test. Indeed, the Gesell test risks a phenomenon of saturation if the subjects tested have surpassed the stage of development it evaluates.

Results

Table 63 presents the results of thyroxine and TSH serum levels and of bone age retardation for all the subjects. The controls demonstrated the characteristic pattern of endemic goitre as compared with the Belgian population, i.e., diminished serum thyroxine and TSH at the upper limit of normal. The myxedematous cretins, as compared with Ubangi controls, had markedly decreased serum T_4 levels and markedly elevated serum TSH levels. These modifications indicated the severity of thyroid insufficiency in this group. The cretinoids had the same abnormalities as the myxedematous cretins but to a lesser degree, except for the serum TSH value, which was increased to the same extent in the two groups. In contrast, the neurologic cretins had serum T_4 values in the same range as controls but the mean value of their serum TSH was significantly higher ($P < 0.001$).

Analysis of the results of serum T_4 and TSH indicated a wide variation in individual values within each group. This variability was particularly evident for serum T_4. In the control group, values characteristic of hypothyroidism (less than 5 μg/dl) were seen in 52 of 83 subjects, in neurologic cretins in 5 of 7. Serum

Table 63. Serum levels of TSH, T_4, and bone age retardation in four groups of subjects.[a]

Group	TSH			T_4			Bone age retardation	
	μU/ml[b]	Number	Variation coefficient	μg/dl (mean ± SE)	Number	Variation coefficient	% (mean ± SE)	Number
Controls	6.7 (6.2–7.3)	79	39	4.7 ± 0.3	83	53	15.6 ± 3.9	7
Cretinoids	113.9*** (84.5–153.6)	12	22	2.6 ± 0.7*	12	91	44.0 ± 12.4*	4
Myxedematous cretins	137.8*** (121.2–156.7)	52	19	1.2 ± 0.3***	52	180	83.9 ± 1.6***	15
Neurologic cretins	36.4*** (13.0–101.7)	7	29	3.6 ± 1.5 NS	7	110	39.0 ± 6.7*	3

[a]Significance: NS = not significant; * = $P < 0.05$; ** = $P < 0.01$; *** = $P < 0.001$.
[b]Geometric mean ($\bar{G}$–1SE, $\bar{G}$+1SE).

TSH values also varied. In six control subjects, values were greater than 50 μU/ml, and in two neurologic cretins values of 50 and 120 μU/ml were observed.

The bone age retardation was significant in cretinoids as compared with controls and even more important in myxedematous cretins. The values observed in neurologic cretins fell between those of controls and those of cretinoids, as they did for the hormone levels.

The results of the Gesell test and Raven coloured matrices are presented in Fig. 75. A progressive decrease in performance was noted from the controls to the neurologic cretins, although there was an important dispersion in individual values with overlaps between the groups. The mean result for the Raven test was very low, even for the controls, and the results for half of them overlapped the results observed for certain myxedematous cretins. Of all the cretins, only two had a normal Gesell test, i.e., a score of 17 arbitrary units, as expected for the age. We found that the mean results were extremely low, taking into consideration the simplicity of the test.

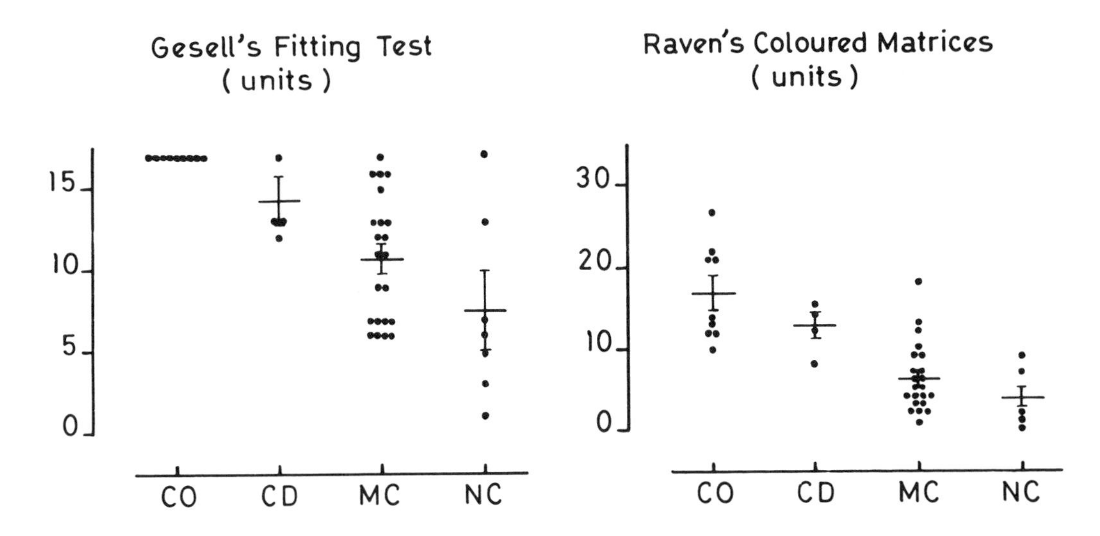

Fig. 75. Results of the intellectual tests in the four groups: controls (CO), cretinoids (CD), myxedematous cretins (MC), and neurologic cretins (NC). The individual values and arithmetic means ± 1 SE are indicated.

Table 64 shows the correlation coefficients between the results obtained in all the controls, the cretinoids, and the myxedematous cretins who took the psychologic tests. The data represent a sample of 35 patients. The correlations between height and the parameters of mental function were highly significant, and particularly significant in relation to the Raven test, which provides a better discrimination between normal and retarded subjects of this age. The table also shows a significant correlation between the results of the psychometric tests and biochemical (T_4 and TSH) or functional (bone age retardation) indices of thyroid function. These correlations were highly significant with the Raven test ($P < 0.001$). The correlations remained significant even when we considered only the control subjects and the cretinoids.

Table 64. Correlation coefficients between different parameters measured in controls, cretinoids, and myxedematous cretins.[a]

Test	Height	Bone age retardation	T_4	TSH (log)
Gesell fitting	0.72 ***	-0.67 ***	0.35 *	-0.52 *
Raven coloured matrices	0.78 ***	-0.71 ***	0.57 ***	-0.66 ***

[a]Significance: NS = not significant; * = P <0.05; ** = P <0.01; *** = P <0.001.

Fig. 76 illustrates the highly significant correlation between the results of the Raven test and the retardation in bone maturation for all the patients, including the two neurologic cretins, in whom both values were known.

Discussion

The definition of endemic cretinism is purely descriptive, as the etiology and pathogenesis of this condition are not yet clear (Querido et al. 1974). The definition is based essentially on mental retardation. As the evaluation of mental development on the basis of clinical observation in mass surveys in the field is very hazardous (Trowbridge 1972; Goslings et al. 1977), only the most extreme forms are reproducibly recognized. The figures for the incidence of cretinism are probably a marked underestimation of the real extent of the syndrome. The present study was designed to quantify intellectual development in a population subjected to extreme iodine deficiency and to seek a relationship between mental development and a series of indicators of thyroid function. The choice of patients was based uniquely on clinical measures of intelligence and thyroid status.

The major contribution of this study was to show that, in patients who did present not the classic stigmata of cretinism but only subtle signs of hypothyroidism and of mental and growth retardation, the results of thyroid activity, intellectual performance, and bone maturation are similar to those found in classic cretinism, only less severe.

This study also demonstrated an important variation in individual results for tests of thyroid function or intellectual performance in each group of subjects, controls, cretinoids, or cretins. This variation suggests that in the groups selected according to clinical criteria there was marked heterogeneity in the adaptation to iodine deficiency and in psychomotor development. A highly significant correlation between the results of thyroid tests and the evaluation of intellectual development was shown, including retardation of growth and bone maturation.

The psychologic tests do not employ language, depending uniquely on the manipulation of geometric forms and abstract drawings. They are generally considered to be culture-free (Geber and Dean 1957; Trowbridge 1972). Their results are not expressed in absolute terms of mental age but rather in relative terms to allow comparison between groups exposed to the same biophysical and cultural environment. These precautions have avoided, at least in part, the biases introduced by many intelligence tests.

Hypothyroidism only causes mental retardation when it occurs early in development and dwarfism when it occurs before termination of growth (Fisher 1976). Thus, the correlations observed between the results of psychologic

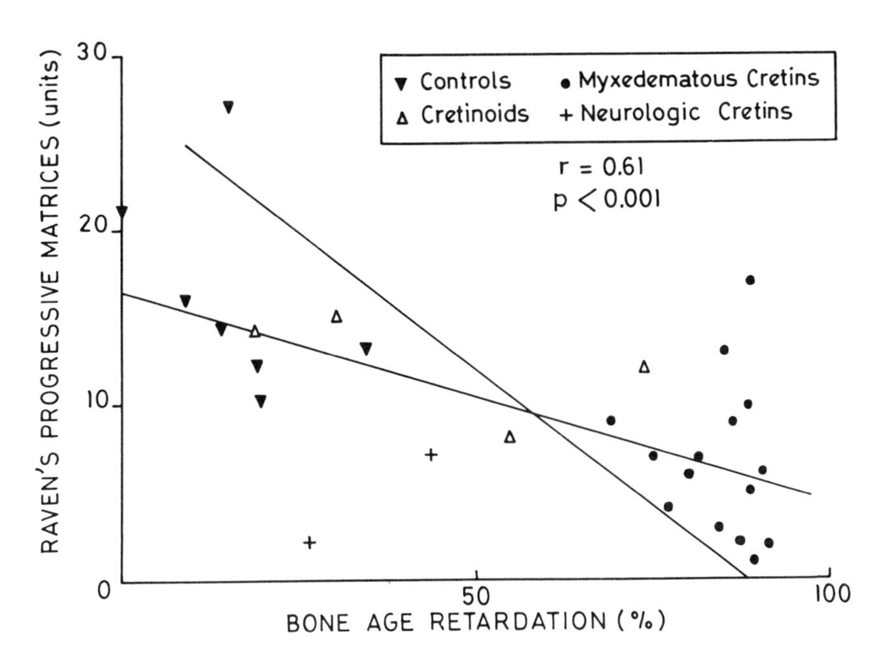

Fig. 76. Scattergram showing the correlation between bone age retardation and Raven test results for all patients in whom both values were known.

tests and tests of thyroid function, in particular bone maturation, strongly support the hypothesis that hypothyroidism occurring early in development is responsible, at least in Africa, for the most frequent form of cretinism, myxedematous cretinism (Dumont et al. 1963; Delange et al. 1972b). This conclusion agrees with previous observations of the high frequency of biochemical signs of thyroid insufficiency at birth in this region (Thilly et al. 1974, 1978). It is in disagreement with the hypothesis of Goslings et al. (1977) who denied the role of hypothyroidism in endemic cretinism in Indonesia because hypothyroid patients were not necessarily mentally retarded. This discrepancy could be explained, at least in part, by a difference in degree of severity due to differences in iodine deficiency (Bourdoux et al. 1978) and in part by the intervention in Africa of natural goitrogens aggravating the effects of iodine deficiency during fetal life (Ermans et al. 1972; Delange 1974; Bourdoux et al. 1978).

The neurologic cretins do not demonstrate the relationship between intellectual and thyroid function observed in other patients. Our results confirmed that they exhibit a marked deficit in intellectual development and a normal thyroid function when compared with controls living in the same region. These observations support the hypothesis of a distinct etiology for this type of cretinism.

In conclusion, this study demonstrated that in severe endemic goitre there is a distinct group of subjects, between normals and cretins, presenting subtle clinical signs that do not correspond to the classic criteria for cretinism but are definite evidence of developmental retardation and hypothyroidism. In addition, our study underlined the important differences between myxedematous and neurologic cretinism from the point of view of the relation between intellectual development and thyroid function. Finally, the study confirmed the existence in populations living in regions of severe endemic goitre of a continuous spectrum of abnormalities of intellectual and physical development and of thyroid function. Cretinism is, therefore, only the extreme expression of the spectrum.

Chapter 13

The Iodine/ Thiocyanate Ratio

F. DELANGE, R. VIGNERI, F. TRIMARCHI, S. FILETTI, V. PEZZINO, S. SQUATRITO, P. BOURDOUX, AND A.M. ERMANS

Thiocyanate potentiates goitre and cretinism among people who are at risk because of lower-than-normal iodine intake. Our studies in goitrous regions of Zaire have confirmed the role of thiocyanate in goitre endemia and have pointed to cassava as the likely mediator. Cassava is a staple of the people's diet and, in many traditional dishes, is a source of considerable cyanide (chapter 6).

Other foods with a high content of thiocyanate-releasing substances potentiate endemic goitre elsewhere (reviewed by Delange and Ermans 1976). For instance, we believe that cabbage is the offender in the diet of Sicilian people who live in one region of endemic goitre.

Our studies in Sicily are noteworthy here because they demonstrated that the balance between iodine and thiocyanate supplies constitutes the crucial factor in the etiology of endemic goitre and cretinism in Sicily and that the balance is more important than iodine supply alone.

Materials and Methods

We studied two goitrous and two nongoitrous areas of Sicily. One goitrous area comprised two communities — Troina (population 13 061, altitude 1121 m above sea level) and Bronte (population 21 600, altitude 760 m above sea level); the other, the village of San Angelo di Brolo (population 5732, altitude 380 m above sea level). These two areas were separated by the Nebrodi Mountains, and there was little contact between the inhabitants. One nongoitrous area was the city of Catania on the east coast, and the other was Messina on the north coast. We chose the two coastal cities as control areas because they were within commuting distance of the goitrous areas but were not goitrous. Although we believed that the inhabitants would be similar in many ways to people living in the goitrous highlands, we also felt that their proximity to the sea meant that they would consume fish and, thus, would have an iodine-rich diet. Therefore, we chose to include another control population whose iodine intake would be at the low end of the normal range but whose thyroid function would be normal — a group from Brussels, Belgium.

Our investigation had two goals:

- To determine the prevalences of goitre and cretinism in the study areas of Sicily and
- To compare the metabolic characteristics of the euthyroid subjects living in goitrous and nongoitrous areas and, thus, to determine differences underlying goitre endemia.

To achieve the first goal, we surveyed male and female schoolchildren, as recommended by the World Health Organization (Perez et al. 1960). We recorded age, sex, and other epidemiological data for computer processing. We examined a total 4427 children aged 5–19, assessing thyroid volume (Perez et al. 1960; Querido et al. 1974) and prevalence of cretinism (Querido et al. 1974).

To reach the second goal, we gathered metabolic data for 436 euthyroid subjects of both sexes, aged 8–71 years: 348 (174 with goitre) from the two endemic areas of Sicily, 24 from Catania, 26 from Messina, and 38 from Brussels, Belgium.

Using 437, 24-h urine samples, we determined daily urinary excretion of iodine and thiocyanate for 342 subjects. We made 1–4 determinations for each subject. We also measured thyroid uptake of radioiodine in 139 subjects (86 from the endemic areas, 21 from Catania, 20 from Messina, and 12 from Brussels). In another 143 nongoitrous subjects (120 from the endemic areas and 23 from Catania), we measured serum TSH, T_3, and T_4 levels.

Results

According to World Health Organization criteria, goitre is endemic if found in 10% of schoolchildren. Our results indicated clearly that goitre was endemic in Troina (55.0%), Bronte (42.0%), and San Angelo (48.7%), whereas Catania (2.2%) and Messina (6.6%) were within the normal range. The percentages of visible goitres and nodules in the endemic areas were respectively 7–40 and 3–10 times as high as in the two control regions (Table 65). Near Bronte, we observed four myxedematous cretins.

The results (Table 66) of our metabolic investigations indicated that only in San Angelo was stable iodine excretion in urine significantly lower than in Brussels ($P < 0.001$); uptake of ^{131}I was correspondingly greater ($P < 0.001$), and thiocyanate levels were not significantly different. Subjects in the other endemic area, Troina and Bronte, however, had ^{131}I uptake and stable iodine excretion values that were not significantly different from controls in Belgium. Notably, they had significantly higher levels ($P < 0.05$ and $P < 0.01$ respectively) of thiocyanate in their urine. In Messina subjects, none of the values were significantly different from those of controls in Brussels. In contrast, the subjects from Catania, which is a sea level urban centre within commuting distance of Troina and Bronte, had values significantly different from those of Belgian controls for all parameters, i.e., ^{131}I uptake was significantly lower ($P < 0.001$); stable iodine in urine was significantly greater ($P < 0.01$) as was thiocyanate ($P < 0.01$).

In the nongoitrous areas — Brussels, Messina, and Catania—the iodine/thiocyanate ratios were similar despite the significant differences in Catania values. Likewise, in the endemic areas, the ratios were similar and much lower than in the control areas.

Serum levels of TSH, T_3, and T_4 (Table 67) were all within the normal range for euthyroid subjects in Catania and the goitrous areas; in particular, the TSH levels were not increased in the endemic areas.

Table 65. Prevalence of visible goitre and nodules among schoolchildren in Sicily.

Place	Subjects	Visible goitre (%)	Nodules (%)
Sea level			
Catania	1188	0.2	1.5
Messina	608	0.7	1.2
Highlands			
Troina	942	8.2	10.6
Bronte	1250	5.1	4.7
San Angelo	439	4.8	12.2

Discussion

Our study confirmed that goitre endemia is severe in the highlands of Sicily. It also showed the differences in etiology of the endemia in the two regions. In one region, San Angelo, goitre is caused by an insufficient dietary intake of iodine, whereas in the other region it is caused by another phenomenon. The iodine excretion in urine of subjects in the Troina–Bronte region was similar to values in Brussels, and, therefore, iodine insufficiency could not account for the serious endemia. Although the iodine excretion in urine was slightly below the normal range, much lower levels have been observed in regions where goitre is not endemic (Roche 1959; Choufoer et al. 1963; Delange et al. 1968). High levels of thio-

Table 66. Levels of thyroidal uptake of radioiodine and daily urinary excretion of iodine and thiocyanate in five centres in Sicily and one in Belgium.[a]

Place	Prevalence of goitre (%)	24-h ^{131}I uptake (% of dose) (mean ± SE)	Urinary excretion: Iodine (μg) (mean ± SE)	Urinary excretion: Thiocyanate (mg) (mean ± SE)	^{127}I/SCN ratio (μg/mg)
Brussels	3.0	49.4 ± 2.3 (12)	51.2 ± 5.8 (38)	6.7 ± 1.3 (17)	7.8
Catania	2.2	31.6 ± 1.5 (21)	113.8 ±-8.4 (79)	13.1 ± 0.9 (85)	8.7
Messina	6.6	50.1 ± 2.1 (20)	76.2 ± 16.6 (9)	8.4 ± 1.9 (9)	9.1
Troina	55.0	50.8 ± 2.4 (28)	40.7 ± 2.6 (90)	11.0 ± 0.7 (103)	3.7
Bronte	42.0	50.8 ± 2.4 (28)	43.2 ± 4.9 (53)	15.2 ± 1.5 (65)	2.8
San Angelo	48.7	70.1 ± 2.5 (30)	26.3 ± 1.7 (130)	7.7 ± 0.3 (137)	3.4

[a]Figures in parentheses are the number of determinations.

cyanate, which has antithyroid properties, were observed in the urine of subjects from Bronte and Troina but were not found in the Belgian controls. Although the levels of thiocyanate were high in the urine of subjects from Catania as well, they were offset by a higher than normal intake of iodine.

Table 67. Serum levels of TSH, T_3, and T_4 in nongoitrous adults in Catania, Troina, Bronte, and San Angelo, Sicily.[a]

Place	TSH (μU/ml) (mean ± SE)	T_3 (ng/100 ml) (mean ± SE)	T_4 (μg/dl) (mean ± SE)
Catania	7.3 ± 0.3 (21)	181 ± 9 (22)	3.5 ± 0.2 (23)
Troina	5.8 ± 0.3 (28)	169 ± 5 (28)	3.7 ± 0.2 (28)
Bronte	8.0 ± 0.4 (32)	153 ± 5 (32)	3.7 ± 0.3 (32)
San Angelo	5.9 ± 0.2 (60)	145 ± 4 (53)	2.7 ± 0.2 (49)

[a]Figures in parentheses are the number of determinations.

What causes the high levels of thiocyanate in the subjects has not yet been determined, although cabbage is the likely mediator. It is consumed in large quantities in Catania, Bronte, and Troina, and it contains thioglucosides (van Etten 1969), which are converted to thiocyanate upon hydrolysis. SCN inhibits iodine accumulation in the thyroid so that its long-term effects are similar to iodine deficiency. The combined effect of high thiocyanate and lower-than-normal iodine levels is similar to the effect of severe iodine insufficiency. The balance between iodine and SCN may be more crucial than the iodine deficiency alone; when the I/SCN ratio falls below a critical level — perhaps 4 — either because of excessively low iodine intake or because of high thiocyanate production, goitre develops.

Chapter 14

General Conclusions

A.M. ERMANS

The epidemiological, clinical, and biologic studies carried out in Kivu and Ubangi document severe goitre endemia, which differs fundamentally from equally severe endemia reported elsewhere because of the very high incidence of myxedematous cretinism. In these areas, the modifications of thyroid metabolism are characteristic of adaptation to a severe iodine deficiency, i.e., elevated uptake of radioactive iodine by the thyroid, decreased serum thyroxine with normal triiodothyronine, and elevated TSH. Iodine deficiency has been demonstrated in measurements of iodine excretion, which only reach levels of 25 μg/24 hours. The role of iodine deficiency has been confirmed by therapeutic trial: administration of iodized oil produces a drastic reduction in goitre prevalence and size, as well as concomitant normalization of parameters of iodine metabolism.

Our observations on the island of Idjwi in Kivu led us to question the classic dogma that iodine deficiency is the only factor responsible for endemic goitre. We observed important differences in goitre prevalence — ranging from 3 to 42% — in different communities on the island, whereas iodine excretion was consistently low in the entire population. This observation was supported by the finding that ^{131}I uptake was abnormally increased in both goitrous and nongoitrous regions. The discrepancy between goitre prevalence and iodine deficiency was first reported by Delange et al. in 1968 and subsequently confirmed by Thilly et al. in 1971. Similar findings had been reported by Roche (1959) in Venezuela and by Choufoer et al. (1963) in New Guinea. On Idjwi, although iodine deficiency could be considered the major causative factor in endemic goitre, an additional goitrogenic factor was obviously intervening and producing important epidemiological variations.

Search for a Goitrogenic Factor

Initially we focused our studies on the goitrogenic properties of certain foodstuffs in humans, paying particular attention to cassava. Earlier, Ekpechi et al. (1966) had established the antithyroid properties of cassava in the rat, and Ermans et al. (1969) had demonstrated that cassava intake was higher in goitrous areas than in nongoitrous ones, although cassava is a mainstay in the diet of all the inhabitants of the island. We tested the influence of the principal foodstuffs consumed in Idjwi on the distribution of a tracer dose of ^{131}I. Following a cassava meal, ^{131}I uptake by the thyroid was clearly reduced, accompanied by increased urinary output of ^{131}I and stable iodine. This finding suggested that cassava contained a thiocyanate-like substance producing a transitory inhibition of the thyroid iodide pump.

As a result of our clinical observations, we undertook experiments in iodine-deficient rats fed fresh cassava roots. Within a short time, we observed a rapid and sustained increase in serum thiocyanate concentrations associated with inhibition of ^{131}I uptake by the thyroid. Thus, the experimental model entirely reproduced the results of the clinical investigation and confirmed that the inhibition of thyroid iodide uptake was due to endogenous production of thiocyanate.

Despite these results, the hypothesis that the goitrogenic action of cassava was uniquely due to repeated inhibition of thyroid iodide uptake was not entirely satisfying. Mean serum thiocyanate concentrations in the inhabitants of Idjwi were elevated, approximately 1 mg/dl as compared with 0.2 mg/dl in Belgians, but these values were not high enough to account for inhibition of the thyroid. In addition, serum thiocyanate levels were equally elevated in non-goitrous and goitrous regions. Therefore, thiocyanate did not appear to be the discriminating factor that we were seeking. Finally, in view of the extremely high ^{131}I uptake observed in the entire Idjwi population, inhibition of uptake, induced by a large cassava meal, probably represented an exceptional rather than normal situation. At this stage, the completion of an iodine prophylaxis campaign rendered further studies impossible on Idjwi.

Before initiating new studies in another goitrous region, we attempted to define more clearly the effects of chronic administration of cassava and thiocyanate in the rat. Data were available on the effects of thiocyanate but mainly on its acute effects. The principal results of this experimental study were as follows:

- Oral administration of thiocyanate (0.2–10 mg/day) did not induce serum thiocyanate levels higher than 1.0–1.2 mg/dl. This same serum concentration was observed following a single injection of 0.2 mg of thiocyanate. Above a critical threshold of 1 mg/dl, a renal adaptive mechanism ensured rapid elimination of the administered doses in the urine. Thus, circulating thiocyanate concentration is not a quantitative measure of thiocyanate load.

- The prolonged administration of high doses of thiocyanate did not cause inhibition of ^{131}I uptake by the thyroid. On the contrary, an increase in uptake was frequently observed.

- Daily ingestion of fresh cassava roots (10 g/day) led to modifications in thiocyanate levels and abnormalities in thyroid iodine metabolism. These modifications were identical, qualitatively and quantitatively, to those induced by 1–2 mg thiocyanate ingested daily. Thus, the antithyroid effect of cassava is directly linked to endogenous thiocyanate production following cassava ingestion.

- The abnormalities in thyroid metabolism induced by cassava or thiocyanate ingestion were characterized by decreased thyroid iodine reserves and the resulting alterations in intrathyroidal hormone production (increased MIT/DIT ratio and decreased T_4 synthesis), in peripheral distribution of thyroid hormones (reduction of $PB^{127}I$), and in thyroid size (goitrogenic action). These abnormalities are identical to those caused by an increase in iodine deficiency.

- In the absence of any inhibition of thyroid iodine uptake, the underlying cause of iodine wastage was not clear. Increased renal iodine clearances could be demonstrated but only for thiocyanate doses of 5 mg/day or higher.

The methods of preparing cassava in Kivu and Ubangi result in partial or complete denaturation of the enzyme linamarase. This enzyme hydrolyzes linamarin, resulting in production of cyanide. With the aim of determining the effect of linamarin in the absence of its specific enzyme, we administered synthetic linamarin to rats in the presence or absence of the enzyme. In its presence, we observed the expected elevation of serum and urinary thiocyanate. In its absence these effects were not seen — a finding that suggests linamarin hydrolysis only occurs when substrate and enzyme are simultaneously present. However, in addition, we tested the hypothesis that linamarin could be degraded nonspecifically by intestinal bacteria. *Klebsiella* were found to degrade linamarin completely. The possible role of intestinal breakdown of linamarin in cassava toxicity in humans has not yet been determined.

Demonstration of a Goitrogenic Effect of Cassava

Despite the lack of definitive evidence of the causative role of cassava in endemic goitre, our findings favoured this hypothesis. Therefore, we initiated clinical studies in Ubangi in northwest Zaire. Endemic goitre in this population of more than 2 million people is even more widespread than that on Idjwi. Cassava intake is also very high. In the goitrous population, mean serum thiocyanate concentrations of 1.0 mg/dl are observed, and some subjects have serum values of 2–3 mg/dl, i.e., 10 times the normal value.

Plasma and urinary thiocyanate concentrations were compared in a large number of subjects. The concentrations increased in parallel up to a serum level of 1.0–1.2 mg/dl, but a renal adaptive mechanism prevented the rise of serum thiocyanate concentration beyond this critical level.

We tested different cassava preparations, calculating their cyanide content and their effects on thiocyanate load in humans. Ingestion of important amounts of cassava for 3 days produced a significant increase in serum and urinary thiocyanate concentrations. Complete cessation of cassava intake induced a clear decrease in these values. Thus, our observations confirmed that cassava is responsible for the abnormal thiocyanate load in this population.

Particularly favourable conditions allowed us to demonstrate the effect of cassava on thyroid metabolism. We were able to investigate a group of adolescents at boarding school. Their food was prepared from local sources but contained less cassava than that of the general population. In addition, the methods of preparation were quite different because of the special kitchen equipment adapted to a large group. Compared with a similar ethnic group living in a rural environment, the boarding school group had lower serum and urinary thiocyanate concentrations. Urinary iodide excretion and ^{131}I thyroid uptake were identical. These data indicate that the two groups were subject to the same degree of iodine deficiency, but the students at boarding school had practically normal serum T_4 and TSH levels, whereas those in the rural environment had decreased T_4 and increased TSH values. These observations clearly suggest that severe iodine deficiency triggers adaptive mechanisms to ensure nearly normal thyroid hormone production. However, the added hindrance from thiocyanate compromises thyroid adaptation and diminishes thyroid secretory capacity even in the presence of intense TSH stimulation.

The goitrogenic action of cassava was also demonstrated in a comparative study of cassava intake and preparation methods in the five different ethnic populations of Ubangi. In three, the Ngbaka, Ngbaka-Mabo, and Mbanza (group I), cassava is consumed principally in the form of *fuku* and *mpondu*. *Fuku* is a gruel of cassava flour mixed with corn. *Mpondu* consists of ground boiled cassava leaves. These two preparations contain an average of 13.5 and 8.5 mg of cyanide per kilogram respectively. The Mongwandi (group II) and the "Gens d'eau" (group III) consume cassava mainly as *chickwangue*, which is soaked and then mashed into a purée. This preparation contains an average 3.5 mg of cyanide per kilogram. The "Gens d'eau" eat large quantities of fish and thus have a higher iodine intake than the rest of the Ubangi population. They are not affected by severe endemic goitre.

Serum and urinary thiocyanate levels in populations consuming *fuku* and *mpondu* (group I) are 1.5- to 2- fold higher than in populations consuming *chickwangue*. There is a highly significant correlation between *fuku* and *mpondu* intake and the prevalence of goitre and cretinism. In groups I and II iodine intake is identical, but the prevalence of goitre is 64% in group I and only 34% in group II. This difference is associated with thiocyanate excretions of 2.57 mg/dl in group I and 1.32 mg/dl in group II. Urinary thiocyanate concentrations are similar in groups II and III, although the "Gens d'eau" (group III) have a lower prevalence of goitre (17%) due to their higher iodine intake.

These results clearly indicate that the prevalence of goitre is linked to the balance between thiocyanate and iodine intake. In each community, prevalence is directly proportional to thiocyanate load and inversely proportional to iodine supply. On the basis of this relationship we have calculated an index corresponding to the ratio between urinary thiocyanate and iodine concentrations. The indices are 2.1 (group I), 0.75 (group II), and 0.39 (group III), corresponding to goitre prevalence of 60, 34, and 17% respectively. Our studies in Sicily (chapter 13) have confirmed the importance of the balance between thiocyanate and iodine supply. Here, iodine intake was at the lower limit of normal, but urinary thiocyanate concentrations were markedly elevated, apparently because of cabbage intake.

No inhibition of ^{131}I thyroid uptake was demonstrable in the clinical situation nor in long-term experiments. Therefore, the exact nature of the antithyroid action of thiocyanate, and consequently of cassava, remained unclear. To clarify this problem, we administered small amounts of thiocyanate, 0.25 mg per day, to iodine-deficient rats, which resulted in serum thiocyanate concentrations of the order of 1.8–1.0 mg/dl. The control group was fed an identical iodine-deficient diet without thiocyanate. ^{131}I uptake was identical in the two groups; however, thyroid uptake of ^{99m}Tc was reduced 50–60% in the rats receiving thiocyanate as compared with the controls. Technetium ion is concentrated in the thyroid but is not organified. A decreased ^{99m}Tc uptake indicates accelerated discharge of technetium from the gland. An identical effect on iodine exit has been demonstrated in rats in which organification is blocked by previous administration of propylthiouracil. These results confirm the hypothesis of Wollman (1962) and Scranton et al. (1969) that small doses of thiocyanate selectively affect the rate of iodine discharge without influencing the concentrating capacity of the thyroid.

It was important to determine why iodine uptake was unchanged despite its rapid exit from the thyroid. One could postulate that when the iodine organification mechanism is intact, intrathyroidal iodide concentration is negligible. Under these conditions, even a very rapid iodine exit would have no significant effect on iodine uptake. However, this explanation was not supported by experiments in which thiocyanate administration was abruptly interrupted. Serum thiocyanate concentrations decreased rapidly, and ^{99m}Tc uptake returned to normal, but ^{131}I was strikingly increased. These observations demonstrate that if ^{131}I uptake apparently remains unchanged in the presence of thiocyanate overload, this value represents a new equilibrium between increased iodine uptake and accelerated iodine exit from the thyroid. This new equilibrium is only attained at the price of increased TSH stimulation of the gland.

More precise quantitative studies should provide an answer to the question of why iodine reserves in the thyroid continue to diminish despite the new equilibrium. In any event, our observations demonstrate that the function of the iodide pump is clearly altered when serum thiocyanate concentrations reach 1.0 mg/dl. The function of the pump can even be disrupted at far lower thiocyanate concentrations that barely exceed physiologic levels.

Another mechanism that could explain the experimental data was recently proposed by van Middlesworth (personal communication). He has observed that thiocyanate overload in the rat induces formation of an abnormal thyroglobulin with sequestration of important amounts of thyroidal iodine. Whatever the underlying mechanism, our observations account for thiocyanate-induced abnormalities in thyroid metabolism in the absence of a decrease in radioactive iodine uptake.

Congenital Hypothyroidism and the Role of Cassava

Approximately 15% of Ubangi newborns present marked hypothyroidism. The deficit in thyroid function usually persists for several months and is, in all likelihood, directly responsible for the developmental abnormalities and mental retardation characteristic of endemic cretinism. The pathogenesis of this syndrome is linked to iodine deficiency, as administration of iodine to the population eliminates congenital hypothyroidism and cretinism. Myxedematous cretinism is only rarely encountered in endemic goitre elsewhere in the world, even in areas where iodine deficiency is as severe as that in Ubangi. The question is whether the incidence of myxedematous cretinism is also influenced by cassava ingestion.

In Ubangi, elevated serum thiocyanate concentrations were observed in pregnant and nursing women as well as in newborns and breast-fed infants. Thiocyanate concentration in cord blood was strictly proportional to the maternal serum concentration — a finding that proves the transplacental passage of this ion. In contrast, thiocyanate content of milk was low and was not influenced by the elevated serum level in nursing mothers. The elevated thiocyanate levels in breast-fed infants are probably due to a supplement of cassava juice, which is frequently given to babies. Thiocyanate ion may have a direct deleterious effect on the thyroid of the fetus and newborn mediated by the same mechanism proposed for adults. In addition, thiocyanate inhibits active transport of iodide in the placenta and mammary glands.

We constructed an experimental model in which pregnant rats were fed an iodine-poor diet supplemented with propylthiouracil and important doses of thiocyanate. Marked thyroid hyperplasia was observed in the newborns but became even more marked on the 16th day of life, i.e., at weaning. Administration of supplementary iodide to the mother 2 days before delivery induced nearly complete normalization of thyroid size in the newborns: however, iodide supplementation during lactation did not produce this effect. These observations suggest that thiocyanate inhibition of the iodide pump in the mammary glands could be an important mechanism in the development of thyroid abnormalities in nurslings.

The Public Health Problem

Epidemiological surveys in Ubangi have revealed that endemic goitre is only an indicator of an overall public health problem of unsuspected enormity. Iodine deficiency together with ingestion of goitrogenic substances produces not only goitre and abnormal thyroid metabolism but also severe abnormalities in pregnant women and in their offspring, particularly during fetal and early life. The following observations document the importance of the problem:

- The incidence of endemic cretinism, essentially myxedematous, is very high, ranging from 1 to 8%. In addition, numerous children are "cretinoid," exhibiting clinical, neurologic abnormalities similar to, but less striking than, cretins.
- Congenital hypothyroidism is present in 15% of newborns. This incidence is 500 times that observed in industrialized countries.
- Birthweight is reduced by 200 grams. This reduction is as marked as that seen in protein-calorie malnutrition, which is known to be associated with a high frequency of malformations and neurologic abnormalities.
- Mortality from birth to 24 months is increased by 7%.
- In Ubangi, psychomotor retardation during the first 2 years of life is demonstrable in infants without outward signs of cretinism. This observation is in keeping with the findings of Fierro-Benitez et al. (1974a) in older children in Ecuador. The concept of widespread psychomotor deficit in children living in goitrous zones appears to be confirmed.

These observations clearly show that severe endemic goitre is an exceptionally important public health problem and an impediment to social and economic development. It is noteworthy that the existence of endemic goitre in the more than 2 million inhabitants of Ubangi was completely unknown until 1973. It is probable that this zone extends well beyond the borders of Zaire. In Africa, as elsewhere in the world, many goitrous zones are still to be discovered. Endemic goitre attracts little interest, probably because it is not spectacular and because its implications are only revealed by long-term epidemiological and biologic investigations.

Prevention of Endemic Goitre and Cretinism

In regions where salt intake is small or variable, iodized salt prophylaxis has little chance of succeeding. In Third World countries, the lack of central facilities to distribute salt and the tropical climate are other reasons for failure of this mode of prophylaxis. The systematic administration of slowly absorbable iodized oil to whole populations was first tested in a pilot project on Idjwi and then administered on a large scale in Ubangi. This method of prophylaxis was effective, cheap, and well accepted. The population was protected for 3–7 years, depending on age and sex. No cases of secondary hyperthyroidism were observed. No adverse effects were seen in pregnant women to whom iodized oil was given at least a month before delivery. Iodized oil administration produced a spectacular decrease in the prevalence and size of goitres. All parameters of thyroid function were corrected for a period longer than 5 years. At birth, children of treated mothers showed normal thyroid function, which persisted for more than 6 months. No cases of cretinism were seen among the offspring of mothers who had received iodized oil.

The cost of the program was approximately U.S. $0.20 per person per year of protection. The injections were carried out by mobile units administering 500–1000 injections per day and averaging 85% coverage of the population. Such a campaign can be organized with moderate means in regions of low socioeconomic development; it has only two prerequisites. One is that the populations must be sensitized to the gravity of endemic goitre through close collaboration between medical personnel and local administrative authorities, and the other is that the program must be carefully planned and monitored with systematic evaluation of the results in samples taken at random. Such control requires specially trained medical and nursing personnel with careful records kept by a central programing organization.

In conclusion, the principal contribution of the clinical and experimental studies described in this monograph is the demonstration that cassava ingestion, together with iodine deficiency, is a key factor in the etiology of endemic goitre and cretinism in central Africa. Thiocyanate is the goitrogenic factor directly involved. The principal difficulty encountered in our work was to distinguish the goitrogenic effects of thiocyanate from those of iodine deficiency. A chronic thiocyanate overload induces abnormalities in thyroid metabolism identical to those observed in iodine deficiency. The observed abnormalities in thyroid function are directly related to the ratio of thiocyanate to iodine.

Cassava is and will continue to be a primary source of nutrition in the Third World. It is, therefore, important to determine iodine supply and goitre prevalence in all regions of high cassava intake. In addition, the necessity of reducing the toxicity of cassava preparations must be emphasized. Two approaches are needed. The populations at risk should be shown how to prepare cassava in a way that ensures complete liberation of hydrocyanic acid, and agricultural researchers should develop subspecies of cassava that have minimal concentrations of linamarin. Also, further research is needed to establish more precisely the thiocyanate/iodine ratio at which prophylactic measures are essential.

Conclusions générales

A.M. Ermans

Les observations épidémiologiques, cliniques et biologiques effectuées au Kivu et dans l'Ubangi mettent en évidence des endémies goitreuses d'une sévérité exceptionnelle. Ces endémies se distinguent exclusivement des endémies d'une sévérité similaire rapportées dans d'autres régions du monde par une prévalence très élevée du crétinisme myxœdémateux.

On retrouve dans ces endémies africaines l'ensemble des modifications du métabolisme thyroïdien caractéristiques d'une adaptation de la thyroïde à une sévère déficience de l'apport iodé, c'est-à-dire une avidité extrême de la glande pour l'iode radioactif, un abaissement du taux sérique de la thyroxine contrastant avec un taux inchangé de triiodothyronine et une élévation du taux de la TSH sérique. La preuve de cette déficience iodée est fournie par des mesures d'excrétion urinaire d'iode qui n'atteignent que 12 à 25 μg/24 h. Le rôle de la carence iodée est confirmé par le test thérapeutique : l'administration d'huile iodée entraîne une réduction drastique de la prévalence et de la taille des goitres ainsi qu'une normalisation concomitante de tous les paramètres du métabolisme iodé.

Nos observations au Kivu sur l'île d'Idjwi nous ont cependant amenés à mettre en doute le dogme classique selon lequel la carence iodée constituerait le seul facteur responsable du goitre endémique. Dans cette île, on observait en effet une disparité extrême de la prévalence du goitre qui variait de 3 à 62 % suivant les communautés. Or, les mesures d'excrétion d'iode urinaire étaient remarquablement superposables dans les communautés goitreuses et non goitreuses, suggérant qu'une carence iodée importante et uniforme atteignait l'ensemble de la population de l'île. Cette constatation était étayée par le fait que dans la zone goitreuse, la captation thyroïdienne d'I^{131} était également anormalement élevée. Cette discordance entre la prévalence du goitre et l'apport iodé fut rapportée d'abord par Delange et al. en 1968; elle fut confirmée ultérieurement par Thilly et al. en 1972 à la suite d'un recensement systématique de l'île. Ces observations inattendues recoupaient d'anciennes données rapportées au Vénézuela par Roche (1959) et en Nouvelle-Guinée par Choufoer et al. (1963). Elles suggéraient que si la carence iodée constituait indubitablement le facteur causal principal de la maladie, un facteur goitrigène supplémentaire devait nécessairement intervenir pour rendre compte des importantes variations épidémiologiques décrites au sein de l'île d'Idjwi.

Recherche d'un facteur goitrigène : mise en cause du manioc

Les recherches effectuées ont été en premier lieu centrées sur les propriétés goitrigènes de certains aliments chez l'homme. Une attention particulière a été portée à la consommation de manioc. Des études antérieures d'Ekpechi et al. (1966) avaient en effet établi les propriétés antithyroïdiennes du manioc chez le rat; or le manioc constituait un des aliments de base de l'alimentation des populations de l'île d'Idjwi. De plus, la consommation de cet aliment était plus importante dans la zone goitreuse que dans la zone non goitreuse (Ermans et al. 1969). Nous avons testé l'influence des principaux aliments consommés à Idjwi sur la distribution d'une dose traceuse d'I^{131}. Il a été ainsi montré que l'ingestion d'un important repas de manioc entraînait une réduction hautement significative de la fixation thyroïdienne de l'iode 131 en même temps qu'une augmentation importante de l'excrétion urinaire d'I^{131} et d'iode stable. Ces observations suggéraient que l'inhibition transitoire de la pompe à iodure thyroïdienne provoquée par le manioc était due à la présence d'une substance du type thiocyanate.

Nous avons transposé ces investigations cliniques à des études expérimentales chez le rat. Des rats carencés en iode ont été nourris avec des carottes fraîches de manioc. Dans un délai très court, on observait une élévation rapide et prolongée de la concentration du thiocyanate sanguin associé à une importante inhibition de la captation d'I^{131} par la thyroïde. Le modèle expérimental reproduisait donc entièrement l'investigation clinique et confirmait que l'inhibition de l'incorporation d'iode par la glande était due à la production endogène de thiocyanate.

En dépit de l'intérêt de ces investigations, la notion que l'action goitrogène du manioc était uniquement conditionnée par des phases répétées d'inhibition de la captation thyroïdienne secondaire à l'ingestion de manioc ne semblait pas entièrement satisfaisante. Le concentration moyenne du thiocyanate sérique des habitants de l'île d'Idjwi était élevée, de l'ordre de 1 mg/dl contre 0,2 mg/dl chez des contrôles belges, mais ces taux n'atteignaient pas les valeurs susceptibles d'inhiber la captation thyroïdienne. De plus, les taux sériques en thiocyanate étaient aussi élevés dans la zone goitreuse que dans la zone non goitreuse. De ce fait, le SCN ne semblait pas constituer le facteur discriminant que nous recherchions. Enfin, les taux de captation extrêmement élevés observés dans l'ensemble de la population de l'île d'Idjwi suggéraient que l'inhibition de la captation d'I^{131} mise en évidence après l'ingestion d'un important repas de manioc constituait une situation probablement exceptionnelle. A ce stade de nos investigations cliniques à Idjwi, la campagne de prophylaxie entamée était achevée, la poursuite de notre étude concernant ce problème y devenait dès lors impossible.

Avant d'entamer de nouvelles investigations dans une autre région goitreuse non traitée, nous avons entrepris de mieux définir l'action du manioc et du thiocyanate dans des conditions d'administration chronique chez le rat. Pour le thiocyanate en particulier, de nombreuses données étaient disponibles mais ne concernaient pour la plupart que son action aigüe.

Les principaux résultats de cette étude expérimentale ont été les suivants:

- Quelles que soient les doses utilisées (0,2 à 10 mg/jour), l'administration orale de thiocyanate pendant plusieurs semaines était incapable d'augmenter la concentration sérique de thiocyanate au delà de 1,0 à 1,2 mg/dl. Cette même concentration sérique était observée déjà après une injection unique de 0,2 mg de SCN en expérimentation aigüe. A partir du seuil critique de 1 mg/dl, un mécanisme d'adaptation rénale assurait l'excrétion très rapide dans les urines de l'entièreté des doses de SCN administrées. La concentration sanguine du thiocyanate ne constituait donc en aucune façon un indice quantitatif de l'apport en thiocyanate.
- L'administration prolongée de ces mêmes doses élevées de thiocyanate n'entraînait aucune inhibition de la captation thyroïdienne de l'I^{131}; au contraire, la captation était fréquemment augmentée dans ces conditions expérimentales.
- L'ingestion journalière de carottes fraîches de manioc (10 g/jour) entraînait simultanément des modifications de la cinétique du thiocyanate et des anomalies du métabolisme thyroïdien. Ces modifications étaient aux points de vue qualitatif et quantitatif exactement superposables aux modifications du métabolisme du thiocyanate et de l'iode thyroïdien induites par la prise journalière de 1 à 2 mg de thiocyanate. L'action antithyroïdienne induite par le manioc était donc bien directement liée à la production endogène de thiocyanate secondaire à l'ingestion de cet aliment.
- Les anomalies du métabolisme thyroïdien provoquées par l'ingestion du manioc et de thiocyanate étaient essentiellement caractérisées par un appauvrissement des réserves iodées de la glande et par ses conséquences classiques sur l'hormonogenèse intrathyroïdienne (augmentation du rapport MIT/DIT et réduction de la synthèse de T_4), sur la distribution des hormones thyroïdiennes à la périphérie

(réduction du PBI^{127}) et sur la taille de la thyroïde (action goitrigène). Ces anomalies étaient donc identiques à celles provoquées par une carence iodée d'une sévérité accrue.

- En l'absence de toute réduction de la captation d'iode par la thyroïde, la cause de la déperdition d'iode observée dans ces conditions expérimentales n'était pas clairement mise en évidence. Une augmentation de la clearance rénale d'iode avait été observée mais uniquement pour des doses extrêmement élevées de thiocyanate (5 mg/jour).

Les procédés de détoxification utilisés par les populations du Kivu et de l'Ubangi pour la préparation du manioc ont pour effet la dénaturation de la linamarase. Cet enzyme hydrolyse la linamarine en cyanure. La question se posait de connaître l'effet de la linamarine lorsqu'elle est ingérée en l'absence de son enzyme spécifique. Dans ce but, de la linamarine synthétique a été administrée à des rats en présence ou en l'absence de linamarase. En présence de linamarase, on observait l'élévation attendue du thiocyanate sérique et urinaire; en l'absence de cet enzyme, ces effets n'étaient pas observés suggérant que l'hydrolyse de la linamarine ne se produit que lors de la présence simultanée du glucoside et de son enzyme spécifique. Nous avons cependant testé l'hypothèse d'une dégradation non spécifique de la linamarine par des bactéries intestinales. Nous avons observé que les *Klebsiella* produisaient une hydrolyse complète de la linamarine. Le rôle éventuel joué par ce mécanisme dans la toxicité du manioc chez l'homme reste à élucider.

Mise en évidence de l'action goitrigène du manioc

Malgré les lacunes qui persistaient dans la démonstration du rôle joué par le manioc dans la pathogénie du goitre endémique, le faisceau d'arguments que nous venons de rapporter rendait l'hypothèse plausible. Nous avons dès lors poursuivi les études cliniques sur le terrain dans l'endémie goitreuse de l'Ubangi au nord-ouest du Zaïre. Cette endémie qui atteint plus de 2 millions d'habitants présente un degré de sévérité nettement supérieur à celui de l'endémie d'Idjwi; elle s'en distingue également par son étendue et son homogénéité. La consommation de manioc y est également très importante. Au sein de la population goitreuse, la concentration sérique en thiocyanate est en moyenne de 1,0 mg/dl; chez certains sujets, on observe des valeurs atteignant 2 à 3 mg/dl c'est-à-dire des valeurs 10 fois supérieures aux valeurs physiologiques.

Une comparaison des concentrations plasmatiques et urinaires en thiocyanate a été effectuée sur un grand nombre de sujets. On y a retrouvé la même relation que celle observée chez les rats : la concentration sérique s'élève proportionnellement à la concentration urinaire jusqu'à une concentration de 1 à 1,2 mg/dl puis s'infléchit et atteint un plateau. Chez l'homme comme chez le rat, un mécanisme d'adaptation rénale empêche donc toute élévation de la concentration sanguine de cet ion, au delà d'un niveau critique.

Nous avons en outre confirmé le rôle par l'ingestion de différentes préparations importantes de manioc sur l'apport en thiocyanate. L'absorption de quantités importantes de manioc pendant une période de trois jours entraîne une augmentation significative des concentrations plasmatiques et urinaires de thiocyanate; inversément, l'interruption complète de cet aliment entraîne une chute significative de ces concentrations. Nos observations confirmaient donc que le manioc était responsable de l'apport anormal en thiocyanate observé dans cette population et qu'il était le principal aliment conditionnant cet effet.

Des circonstances particulièrement favorables nous ont permis d'étayer l'influence de l'ingestion du manioc sur le métabolisme thyroïdien. Nous avons en effet pu suivre au sein de cette population un groupe de jeunes filles habitant un pensionnat dont l'alimentation est assurée par des produits locaux mais où les quantités de manioc consommées sont notablement plus basses que dans la population générale. De plus, les conditions de préparation du manioc sont sensiblement différentes vu l'utilisation d'un matériel de cuisine spécialement adapté aux besoins d'une grande communauté. Par rapport à un groupe de jeunes filles de la même ethnie mais vivant en milieu rural, les jeunes filles du pensionnat ont des taux de thiocyanate sériques et urinaires abaissés, alors que l'excrétion de l'iode urinaire et les taux de captation d'I^{131} sont superposables. Ces données indiquaient que les deux groupes examinés étaient soumis à une carence iodée identique. Les jeunes filles du pensionnat présentaient des taux sériques de T_4 et de TSH pratiquement normaux alors que ces taux étaient fortement abaissés pour la T_4 et augmentés pour la TSH chez les jeunes filles vivant en milieu rural. Ces observations indiquent clairement que dans une situation de carence iodée sévère, les mécanismes d'adaptation permettent de maintenir une sécrétion thyroïdienne presque normale. Par contre, l'association à cette carence iodée d'une élévation du taux sanguin en thiocyanate compromet l'efficacité de ces processus d'adaptation et réduit considérablement la capacité de la thyroïde à secréter des quantités adéquates de thyroxine malgré un degré notablement accru de sa stimulation par la TSH.

L'action goitrigène du manioc a également été mise en évidence grâce à une étude comparative de la consommation de cet aliment et de son mode de préparation dans les différentes ethnies constituant la population de l'Ubangi. Le manioc est le principal aliment de base dans les six ethnies dénombrées dans la zone étudiée. Dans trois d'entre elles, chez les Ngbaka, les Ngbaka-Mabo et les Mbanza (groupe I), le manioc est surtout consommé sous forme de *fuku* et de *mpondu*. Le *fuku* est une pâte de manioc non roui agglomérée avec du maïs; le *mpondu* est constitué de feuilles de manioc bouillies et broyées. Ces deux aliments contiennent en moyenne respectivement 13,5 et 8,5 mg de cyanure par kilo. Dans les deux autres ethnies, les Mongwandi (groupe II) et chez les Gens d'eau (groupe III), le manioc est essentiellement consommé sous forme de *chickwangue*, c'est-à-dire une purée de manioc roui contenant en moyenne 3,5 mg de cyanure/kilo. Les Gens d'eau, grands consommateurs de poissons, ont un apport iodé nettement plus élevé que le reste de la population de l'Ubangi; cette ethnie est située en dehors de la zone d'hyperendémie goitreuse.

Les concentrations sanguines et urinaires en thiocyanate des ethnies qui consomment du *fuku* et du *mpondu* (groupe I) sont 1,5 à 2,0 fois plus élevées que les ethnies consommant la *chickwangue*. Une étude statistique montre une corrélation hautement significative entre la consommation du *fuku* et du *mpondu* et la prévalence du goitre et du crétinisme. Dans les ethnies des groupes I et II, pour un apport iodé strictement superposable, les prévalences du goitre sont respectivement de 60 et de 34 %, différence qui est associée à une concentration du thiocyanate urinaire de 2,57 mg/dl pour le groupe I et de 1,32 mg/dl pour le groupe II. A l'inverse, les concentrations urinaires du thiocyanate des groupes II et III sont similaires, mais comme nous venons de le mentionner, les Gens d'eau (groupe III) présentent une prévalence du goitre nettement moins importante (17 %) en rapport avec un apport iodé plus élevé.

Ces observations indiquent clairement que la prévalence du goitre est liée à une balance entre les apports de thiocyanate et d'iode : la prévalence du goitre au sein de chaque communauté croît en fonction directe du degré d'imprégnation en thiocyanate et en fonction inverse de l'apport iodé. Sur la base de cette constatation, nous avons calculé un index correspondant au rapport entre les concentrations du thiocyanate et de l'iode dans les urines. Dans les trois groupes d'ethnies de l'Ubangi que nous avons distingués, ce rapport est de 2,1 pour le groupe I, 0,75 pour le groupe II et 0,39 pour le groupe III. Dans ces trois groupes, les prévalences de goitre sont respectivement de 60, 34 et 17 %. Confirmant l'importance jouée par l'équilibre entre les apports en iode et en thiocyanate, nous avons récemment décrit en Sicile une endémie goitreuse au sein de laquelle l'apport iodé est à la limite inférieure de la normale, mais où les concentrations urinaires du thiocyanate sont très élevées, en rapport, semble-t-il, avec une consommation importante de choux.

Aucune inhibition de la captation thyroïdienne de l'iode n'ayant été mise en évidence dans des conditions cliniques ou expérimentales chroniques, la nature exacte de l'action antithyroïdienne du thiocyanate et a fortiori du manioc restait à expliquer. Pour tenter d'élaircir ce problème, nous avons chez des rats carencés en iode administré de façon continue de petites doses de thiocyanate (0,25 mg/jour) qui augmentent la concentration sanguine de cet ion à des valeurs de l'ordre de 0,8 à 1,0 mg/dl; le groupe témoin était constitué par des rats recevant uniquement la même diète pauvre en iode. Dans les deux groupes d'animaux, la captation thyroïdienne de l'iode 131 est strictement identique. Simultanément, nous avons pratiqué des mesures de captation thyroïdienne du 99^{m}Tc-pertechnétate, ion qui est concentré comme l'iode par la thyroïde mais qui n'est pas organifié. Chez les rats traités par thiocyanate, la captation thyroïdienne du technétium est réduite de 50 à 60 % par rapport aux rats témoins. Cette constatation témoigne d'une ressortie accélérée du technétium de la glande. Elle est superposable à celle rapportée pour l'iodure chez des rats dont l'organification est bloquée par l'administration préalable de propylthiouracil, elle confirme également l'hypothèse de Wollman (1962) et de Scranton et al. (1969) selon laquelle, à faibles doses, le thiocyanate agit exclusivement sur la vitesse de ressortie de l'iodure libre de la thyroïde sans influencer le processus de concentration de cet ion par la glande.

La question se posait de savoir pourquoi malgré l'importance de cette ressortie d'iodure, la captation thyroïdienne de cet ion restait inchangée.

Une explication plausible consistait à postuler que lorsque le mécanisme d'organification de l'iode est intact, l'efficacité de ce processus est telle que la concentration intrathyroïdienne d'iodure est négligeable. Dans ces conditions, la ressortie d'iodure, même fortement accélérée, pourrait n'avoir qu'un effet négligeable sur la quantité d'iode captée par la glande. Cette explication a cependant été infirmée par des observations réalisées en interrompant brutalement la surcharge en thiocyanate. En effet, la décroissance rapide de la concentration sérique en thiocyanate observée dans ces conditions, est associée non seulement à une normalisation de la captation du 99^{m}Tc mais également à une très importante augmentation de la captation d'I^{131}. Ces observations démontrent que si le niveau de fixation thyroïdienne de l'iode 131 reste apparemment inchangé sous imprégnation chronique de thiocyanate, cette observation résulte en fait de l'établissement d'un nouvel équilibre entre une captation notablement augmentée de l'iode au niveau de la glande et la ressortie accélérée de cet ion induite par le thiocyanate. Ce nouvel équilibre n'est atteint qu'au prix d'un degré de stimulation accrue de la thyroïde par la TSH.

Des études quantitatives plus précises devront permettre de préciser pourquoi malgré cet équilibre, la glande continue à s'appauvrir en iode; nos observations démontrent de toute façon que la pompe à iodure thyroïdienne est nettement perturbée pour des concentrations sériques de thiocyanate de l'ordre de 1,0 mg/dl. Cette anomalie est même détectée à des concentrations beaucoup plus faibles, à peine supérieures aux concentrations physiologiques.

Un autre mécanisme susceptible d'expliquer les données expérimentales a été rapporté récemment par van Middlesworth (communication personnelle). Cet auteur observe en effet qu'une surcharge en thiocyanate chez le rat provoque la formation d'une thyroglobuline anormale entraînant une séquestration de quantités importantes d'iode dans la glande. Quel que soit le mécanisme déterminant, ces observations permettent d'expliquer comment une surcharge chronique en thiocyanate est à même de perturber le métabolisme thyroïdien sans réduire la captation du radioiode.

L'hypothyroïde congénitale et le rôle étiologique du manioc

Environ 15 % des nouveau-nés de l'endémie goitreuse de l'Ubangi présentent une hypothyroïdie très marquée. Ce déficit thyroïdien persiste chez la plupart d'entre eux pendant plusieurs mois et est très probablement directement responsable des anomalies de développement et de l'arriération mentale qui sont caractéristiques du crétinisme endémique. La pathogénie de ce syndrôme est liée à la carence iodée puisque l'administration d'iode dans cette population entraîne une disparition de l'hypothyroïdie congénitale et du crétinisme. Comme nous l'avons déjà mentionné, le crétinisme myxœdémateux n'est observé que de façon exceptionnelle dans les autres endémies goitreuses décrites dans le monde, même dans celles où la sévérité de la carence iodée atteint celle observée dans l'Ubangi. La question se posait donc de savoir si la prévalence particulière de ce syndrôme dans l'Ubangi n'était pas également influencée par l'ingestion du manioc.

Dans l'Ubangi, des concentrations sériques élevées en thiocyanate ont été observées chez les femmes enceintes et au cours de la lactation ainsi que chez les nouveau-nés et les enfants en période d'allaitement. La concentration en thiocyanate au niveau du cordon est strictement proportionnelle à la concentration sérique de la mère, observation qui traduit bien la diffusion transplacentaire de cet ion. Par contre, le contenu en thiocyanate du lait est bas et n'est pas influencé par la concentration élevée du thiocyanate chez les mères allaitantes. Le taux élevé de thiocyanate observé chez les enfants en cours d'allaitement est probablement causé par un supplément de jus de manioc qui est fréquemment donné à ces nourrissons. Une action délétère du thiocyanate est susceptible de s'exercer par une influence directe de cet ion sur la thyroïde du fœtus ou du nouveau-né suivant le même processus que celui décrit chez l'adulte. A ce mécanisme peut s'ajouter une action inhibitrice du thiocyanate sur le transport actif de l'iodure au niveau du placenta et des glandes mammaires.

Un modèle expérimental a été étudié chez le rat en traitant des rattes gravides par une diète pauvre en iode, par du propylthiouracil et d'importantes doses de thiocyanate. Une hyperplasie marquée de la thyroïde a été observée chez les rats nouveau-nés, mais cette hyperplasie est beaucoup plus marquée au 15ᵉ jour de vie c'est-à-dire à la fin de la période de lactation. L'administration d'un supplément d'iode à la mère deux jours avant l'accouchement entraîne une quasi normalisation du poids de la thyroïde des nouveau-nés. Par contre, cette administration reste sans effet lorsque la même injection est effectuée en cours de lactation. Ces observations suggèrent donc que l'inhibition de la pompe à iodure de la glande mammaire par le thiocyanate pourrait constituer un mécanisme important dans le développement des altérations thyroïdiennes observées chez les nourrissons.

Gravité du problème de santé publique

Les enquêtes épidémiologiques réalisées en Ubangi révèlent que le goitre endémique n'est en fait que la marque d'un problème de santé publique d'une gravité insoupçonnée. En effet, en dehors du goitre et d'un fréquent déficit de la fonction thyroïdienne, la carence iodée conjuguée à l'absorption de substances goitrigènes est responsable d'anomalies graves atteignant principalement la femme enceinte et l'enfant surtout pendant la vie fœtale et périnatale. Ces anomalies ont été objectivées par les observations suivantes :

- une incidence très élevée du crétinisme endémique, essentiellement sous la forme de crétinisme myxœdémateux; qui peut atteindre jusqu'à 1 à 8 % de la totalité des communautés, de plus, de nombreux enfants étiquetés "crétinoïdes" présentent des atteintes neurologiques moins évidentes au point de vue clinique mais néanmoins prouvées de façon indiscutable par des tests spécifiques;
- une incidence de l'hypothyroïde congénitale qui atteint 15 % des nouveau-nés, soit 500 fois plus élevée que celle observée dans les pays industrialisés;
- une réduction du poids des enfants à la naissance de près de 200 g, qui est aussi importante que celle observée dans la malnutrition protéico-calorique classiquement associée à une fréquence élevée de malformations diverses et de troubles neurologiques;
- une augmentation de la mortalité infantile de 7 % pendant la période s'étendant depuis la naissance jusqu'au 25ᵉ mois;
- un retard dans le développement psychomoteur, qui, dans l'Ubangi, a pu être établi jusqu'à l'âge de 24 mois chez des enfants ne présentant aucun stigmate de crétinisme. Nos observations recoupent les constatations de Fierro-Benitez et al. (1974a) réalisées en Équateur chez des enfants plus âgés. La notion d'un déficit psychomoteur atteignant une partie importante sinon la totalité des enfants vivant dans les communautés goitreuses semble donc confirmée.

L'ensemble de ces observations met clairement en évidence que le goitre endémique sévère masque en réalité une plaie sociale et médicale d'une gravité exceptionnellement importante au point de vue de la santé publique et constituant très probablement un frein considérable au développement socio-économique de ces populations. A cet égard, il est intéressant de noter que l'existence de l'endémie goitreuse de l'Ubangi qui compte près de 2 millions d'habitants est restée totalement méconnue jusqu'en 1973. Il est probable que les limites de cette endémie débordent largement les frontières du Zaïre et qu'en Afrique comme ailleurs dans le monde, de nombreux foyers de cette maladie restent encore à découvrir. Cette absence d'intérêt pour le goitre endémique réside dans son caractère peu spectaculaire et surtout dans le fait que l'importance de ses complications ne peut être précisée qu'après des études épidémiologiques et biologiques de longue haleine.

La prévention du goitre et du crétinisme

Dans les régions où la consommation du sel est peu importante ou inhomogène, la prophylaxie par sel iodé s'avère illusoire. L'absence de centralisation des circuits de distribution du sel ainsi que le climat tropical sont deux autres causes de l'échec de ce mode de prophylaxie dans les pays du Tiers-Monde. L'injection systématique d'huile iodée lentement résorbable à l'ensemble de la population a été réalisée d'abord dans une étude pilote à Idjwi, puis sur une grande échelle dans l'Ubangi. Cette méthode prophylactique s'est avérée efficace, peu coûteuse et inoffensive; la protection de la population est assurée pendant une durée de 3 à 7 ans suivant l'âge et le sexe des patients. Aucun cas d'hyperthyroïdie secondaire à ce traitement n'a été observé. Une même innocuité de l'huile iodée a été observée lorsque l'injection est réalisée chez la femme enceinte à moins d'un mois de l'accouchement. L'administration d'huile iodée est suivie d'une réduction spectaculaire de la prévalence et de la taille des goitres. L'ensemble des paramètres de la fonction thyroïdienne est corrigé pendant plus de 5 ans. Au moment de la naissance, les enfants de mères traitées présentent une fonction thyroïdienne normale, et cette correction se maintient après plus de 6 mois. Aucun cas de crétinisme n'a été observé chez les mères traitées par cette méthode.

Le coût final du programme est de l'ordre de 0, 2 $ US par habitant et par année de protection. Les injections sont réalisées par des équipes mobiles qui peuvent effectuer 500 à 1 000 injections par jour. Une telle campagne peut être organisée avec des moyens réduits, dans des régions à faible niveau socio-économique. Son intérêt majeur est une correction homogène de la population avec une couverture moyenne de l'ordre de 85 %. Deux conditions déterminent la réussite de tels programmes : l'une est une sensibilisation des populations à la gravité du goitre endémique, subordonnée à une collaboration étroite des autorités médicales et administratives locales. La seconde est une programmation extrêmement précise et un contrôle étroit de la progression de la campagne de traitement ainsi que l'évaluation systématique de son efficacité sur des échantillons de populations sélectionnées au hasard. Ce contrôle ne peut être réalisé que par un personnel médical et infirmier spécialement formés et grâce à l'utilisation d'un système de fiches soigneusement tenues à jour et centralisées par un organisme de programmation unique.

En conclusion, la principale contribution des études cliniques et expérimentales rapportées dans cette monographie est la démonstration que conjointement à la carence iodée, la consommation de manioc intervient dans l'étiologie du goitre endémique et très probablement dans celle du crétinisme endémique en Afrique centrale. La substance goitrigène directement en cause est le thiocyanate. La principale difficulté rencontrée au cours de cette étude a été de distinguer les effets goitrigènes du thiocyanate de ceux de la carence iodée. En effet, une augmentation prolongée de l'apport en thiocyanate entraîne des anomalies du métabolisme thyroïdien en tous points identiques à celles induites par la déficience de l'apport iodé. Le facteur déterminant les anomalies thyroïdiennes est constitué par la balance entre l'apport en iode et l'apport en thiocyanate.

Étant donné l'importance primordiale que joue et qu'est appelé à jouer le manioc dans l'alimentation des pays du Tiers-Monde, nos observations impliquent la nécessité de surveiller l'apport iodé et la prévalence du goitre dans toutes les régions où cet aliment est consommé. Nos constatations impliquent également la nécessité de réduire la toxicité des aliments préparés à partir du manioc. Un tel résultat peut être atteint grâce à deux types de mesures. Sur le plan diététique, il faudrait tenter de faire appliquer par les populations à risques des procédés de préparation du manioc assurant une complète libération de son contenu en acide cyanhydrique. Sur le plan agricole, il y a un intérêt évident à promouvoir l'utilisation systématique de sous-espèces de manioc contenant des concentrations minimales de linamarine. Enfin, nos observations démontrent qu'en présence d'une carence iodée sévère, l'introduction urgente d'une prophylaxie iodée s'impose de façon absolue. Des études ultérieures devraient permettre de préciser à partir de quel degré de déficience de l'apport iodé, ces mesures prophylactiques doivent être appliquées.

Appendix 1

The Gemena Centre

The Gemena central unit, occupying more than 500 m^2 space, was set up in a group of buildings renovated by the IRS, comprising offices for administrative use, a meeting room, a large laboratory, washing facilities, storeroom, library, three apartments, a kitchen, and a workshop.

Since the laboratory's inauguration in 1975, several thousand measurements have been performed there to determine cyanide in foods, serum and urinary thiocyanate concentrations, and urinary creatinine concentration.

The laboratory equipment, financed principally by the IDRC, includes a high-speed centrifuge, two portable centrifuges, a semi-analytic balance, an analytic balance, a spectrophotometer with UV and visible light sources, and an automatic sampling system, a pH meter, an incubator equipped for agitation, a chemical hood, a refrigerator, and a freezer. The unit can operate almost autonomously, as it has a water reservoir and an electric generator. The washing facilities include a system for deionizing water and a large capacity oven. The library has several hundred books and periodicals.

General view of the IRS–CEMUBAC centre in Gemena.

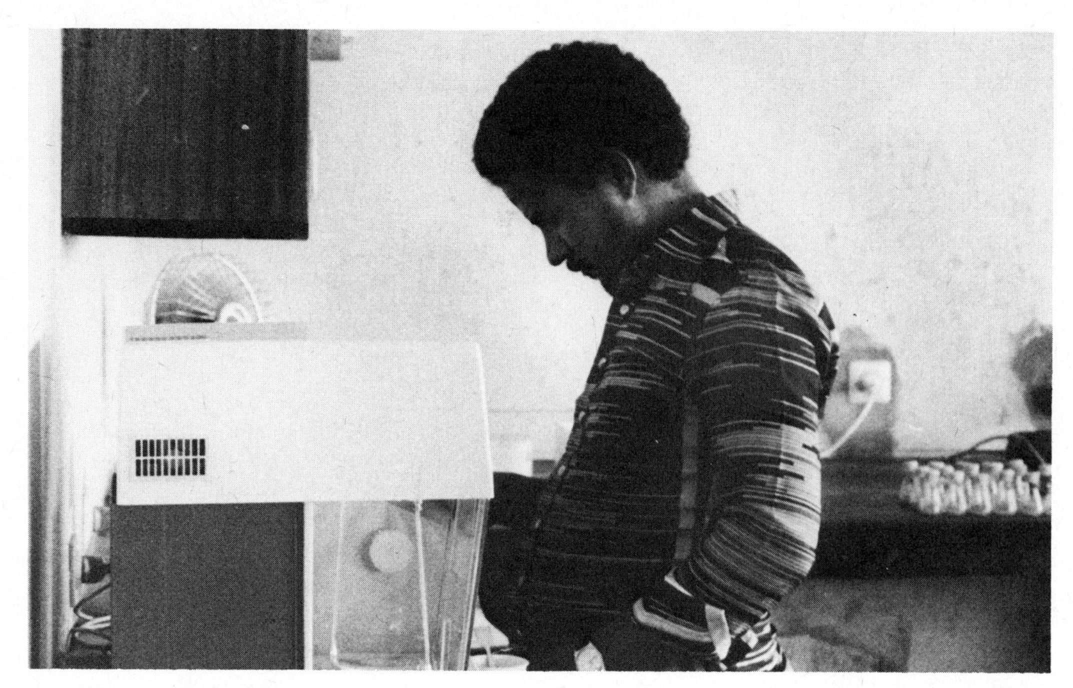

The Zairian chemist at work: weighing food samples for further assessment of their cyanide content.

View of the laboratory with general equipment.

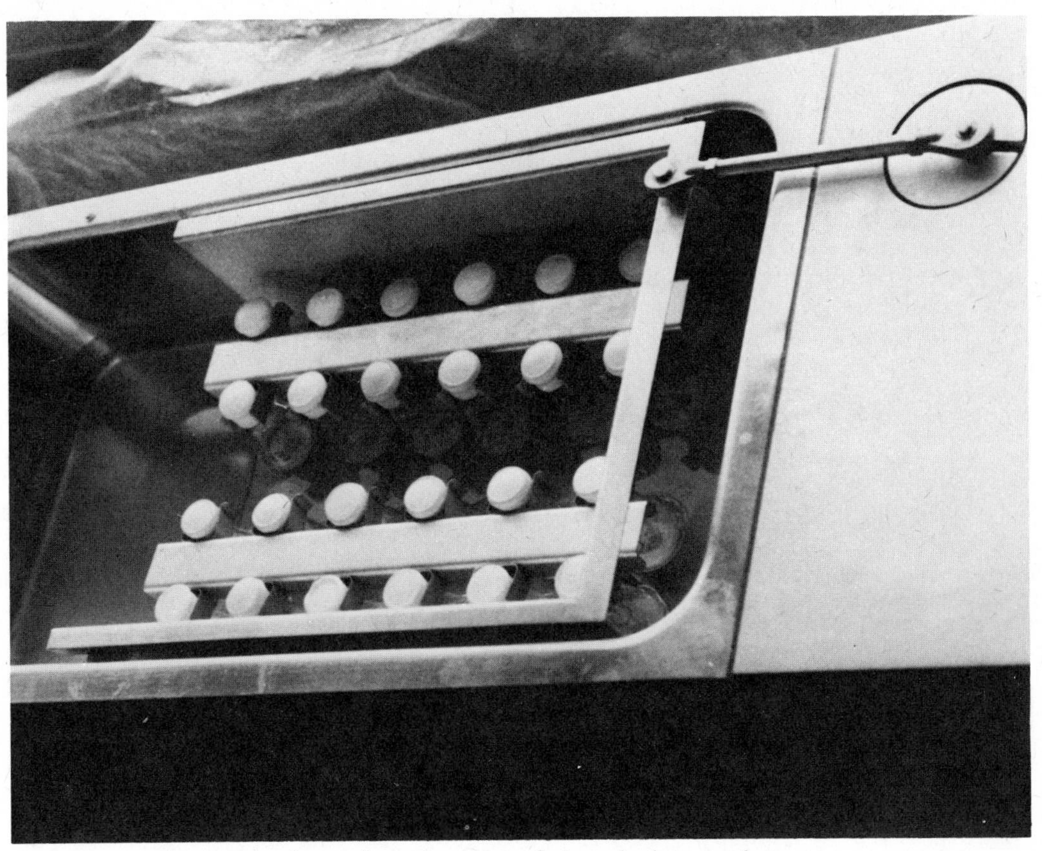

Incubation of cyanide-containing food samples during autolysis.

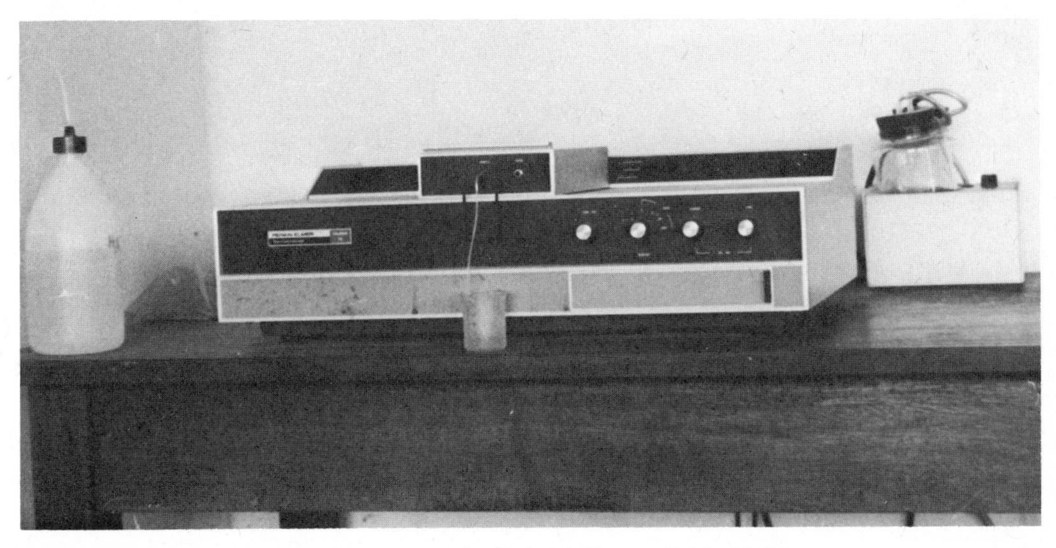

Spectrophotometer: another kind of facility provided by IDRC.

Appendix 2

General Methodology

Human T_4: Our team measured serum T_4 levels in duplicate (or triplicate), using the following radioimmunoassay procedure: to 25 μl serum or standard we added 300 μl of antiserum to T_4 (Henning, Berlin) in barbital buffer (pH 8.5; 0.05M; bovine gammaglobulin 0.75%; ANS 0.03%; Thimerosal 0.01%) at a final dilution of 1/10 500 and 100 μl of $[^{125}I]T_4$ (specific activity > 1200 μCi/μg; the Radiochemical Centre, Amersham). Final dilution of antiserum 1/10 500 ensured a binding of approximately 50% (Bo/T) of the total tracer. The mixture was incubated at room temperature for 2 hours; we separated bound and free fractions by adding 2 ml of 18% PEG 6000 in barbital buffer (pH 8.5; 1.0M; BSA 0.1%). By diluting appropriate amounts of T_4 (0; 1.25; 2.5; 5.0; 10.0; 20.0 μg/dl; RIA standard, Henning, Berlin) in human serum, which had been made iodothyronine-free with NORIT OL 20 g/100 ml serum (Mitsuma et al. 1972), we prepared the standards and ensured quality control using three different sera (low, normal, and high range). For a serum of 5.5 μg/dl the within-assay coefficient of variation was 3.5% and the between-assay coefficient was 5.8%.

Human T_3: Part of the study was performed with T_3 RIA kits (the Radiochemical Centre, Amersham) according to the method described by Hüfner and Hesch (1973). Subsequently, serum T_3 levels were measured in duplicate (or triplicate) with the following RIA: to 50 μl serum or standard were added 200 μl of antiserum to T_3 (Henning, Berlin) in barbital buffer (pH 8.5; 1.0 M; bovine γ-globulin 0.75%; Thimerosal 0.1%) at a final dilution of 1/22 500 and 200 μl of $[^{125}I]$ T_3 (specific activity > 1200 μCi/μg; the Radiochemical Centre, Amersham). Final dilution of antiserum (1/22 500) ensured a binding of approximately 40% (Bo/T) of the total tracer. The mixture was incubated at 37 °C for 2 hours, and adding 2 ml of 18% PEG 6000 in barbital buffer (pH 8.5; 0.09 M; BSA 0.1%) we separated bound and free fractions. Standards were appropriate amounts of T_3 (0, 25, 50, 100, 250, 500 ng/dl; RIA Standard; Henning, Berlin) diluted in human serum made iodothyronine-free (procedure as for T_4). We performed quality control procedures with three different sera (low, normal, and high range). For a serum of 152 ng/dl the within-assay variation coefficient was 4.1% and the between-assay coefficient was 6.9%. Correlation between the two methods gave r = 0.9421 for 459 samples.

Human TSH: Serum TSH levels were determined in duplicate (or triplicate) with h-TSH RIA kits (Abbott Laboratories Diagnostic Division, North Chicago, Illinois) based on the method of Odell et al. (1965a,b). The team performed quality control procedures with two different samples (normal and high range) prepared with h-TSH standard 68/38 kindly provided by the WHO International Laboratory for Biological Standards, Holly Hill, Hampstead, London.

Rat T_4: Serum T_4 levels were measured in triplicate; the following RIA procedure was used: to 25 μl serum or standard were added 300 μl of antiserum to T_4 (Henning, Berlin) in barbital buffer (pH 8.5; 0.05M; bovine γ-globulin 0.75%; ANS 0.03%; Thimerosal 0.01%) at a final dilution of 1/21 000 and 100 μl of $[^{125}I]T_4$ (specific activity > 1200 μCi/μg; the Radiochemical Centre, Amersham). Final dilution of antiserum (1/21 000) ensured a binding of approximately 50% (Bo/T) of the total tracer. The mixture was incubated at room temperature for 3 hours. 100 μl of NRS (2.5% in barbital buffer) and 100 μl of sheep anti-rabbit gammaglobulin at a dilution of 1/5 in barbital buffer (sheep anti-rabbit gammaglobulin kindly provided by Mr Leclercq, Institut Supérieur d'Enseignement Technique de l'État, Irchonwelz, Belgium) were added as precipitating agent, and the total was incubated a further 24 hours at 4 °C. Standards were appropriate amounts of T_4 (0, 0.625, 1.25, 2.5, 5.0, 10.0 μg/dl; RIA standard; Henning, Berlin) diluted in rat serum made iodothyronine-free (procedure as for human T_4).

Rat T_3: Serum T_3 levels were measured in triplicate; the RIA procedure was to begin with 100 μl serum or standard, add 100 μl of antiserum to T_3 (Henning, Berlin) in barbital buffer (pH 8.5; 0.05M; bovine γ-globulin 0.75%) at a final dilution of 1/32 000, 100 μl of barbital buffer (pH 8.5; 0.05M; bovine γ-globulin 0.75%; ANS 0.03%; Thimerosal 0.01%) and 100 μl of $[^{125}I]T_3$ (specific activity 1200 μCi/μg; the Radiochemical Centre, Amersham). Final dilution of antiserum 1/32 000 ensured a binding of approximately 32% (Bo/T) of total tracer. The mixture was incubated at 4 °C for 24 hours; 100 μl of NRS (2.5% in barbital buffer) and 100 μl of sheep anti-rabbit gammaglobulin at a dilution of 1/5 in barbital buffer (sheep anti-rabbit gammaglobulin kindly provided by Mr Leclercq, Institut Supérieur d'Enseignement Technique de l'État, Irchonwelz, Belgium) were added as a precipitating agent and again the mixture was incubated at 4 °C for 24 hours. By diluting appropriate amounts of T_3 (0, 25, 50, 100, 200, 400 ng/dl; RIA standard; Henning, Berlin) in rat serum made iodothyronine-free (procedure as for human T_4), we prepared the standards.

Rat TSH: Serum TSH levels were measured in triplicate, and the following procedure was used: to 100 μl serum or standard were added 100 μl of antiserum to rat-TSH in phosphate buffer (pH 7.6; 0.01M; NaCl 0.15M; EDTA 0.05M; Thimerosal 0.01%; NRS 0.2%) at a final dilution of 1/40 000, 100 μl of tracer and 200 μl of phosphate buffer (pH 7.6; 0.01M; NaCl 0.15M; EDTA 0.05M; Thimerosal 0.01%; BSA 0.2%). Our team obtained tracer with a specific activity 150 μCi/μg, using the iodination procedure of Redshaw and Lynch (1974). The mixture was incubated at 4 °C for 4 days. Separation of bound and free fractions was achieved by the addition of 100 μl NRS (2.5% in phosphate buffer) and 100 μl of sheep anti-rabbit gammaglobulin at a dilution of 1/4 in phosphate buffer (sheep anti-rabbit gammaglobulin kindly provided by Mr Leclercq, Institut Supérieur d'Enseignement Technique de l'État, Irchonwelz, Belgium) and further incubation at 4 °C for 24 hours. We prepared standards by diluting appropriate amounts of rat-TSH (0, 10, 25, 50, 100, 250, 500, 1000 ng/100 μl) in phosphate buffer (pH 7.5; 0.01M; NaCl 0.15M; Merthiolate 0.01%; BSA 1.0%). All reagents for the measurements of rat-TSH were kindly provided by the NIAMDD (National Institute of Arthritis, Metabolism and Digestive Diseases) NIH, Rat pituitary hormone distribution program.

Results

All RIA results were calculated in a logit–log system (Rodbard et al. 1968). *Thiocyanate* in serum, urine, and milk was measured according to the procedure of Aldridge (1945) modified by Michajlovskij and Langer (1958). *Iodide* in urine was determined by the method of Riley and Gochman (1964). *Uptake of radioactive iodine* was determined according to the method described in Bourdoux et al. (1978).

References

Abrol, Y.P., and Conn, E.E. 1966. Studies on cyanide metabolism in *Lotus arabicus* L. and *Lotus tenuis*. Phytochemistry 5, 237–242.

1967. Studies on the biosynthesis of amygdalin, the cyanogenic glucoside of bitter almonds (*Prunus amygdalus* Stokes). Indian J. Biochem. 4, 54–55.

Aldridge, W.N. 1945. The estimation of micro quantities of cyanide and thiocyanate. Analyst (London), 70, 474–475.

Andersen, H.J. 1961. Studies of hypothyroidism in children. Acta Paediatr. 50, (Suppl. 125), 1–150.

Andrews, J. 1973. Thiocyanate and smoking in pregnancy. J. Obstet. Gynaecol. Br. Commw.80, 810–814.

Anonymous. 1969. Chronic cyanide neurotoxicity. Lancet 2, 942–943.

1970. Tobacco amblyopia. Can. Med. Assoc. J. 102, 420.

1977a. Toxicity of laetrile. FDA Drug Bulletin 7, 26–32.

1977b. Laetrile. The political success of a scientific failure. Consumer Reports, August 1977, 444–447.

Aquaron, R. 1977. Urinary, salivary and plasma levels of thiocyanate in goitrous and non goitrous areas of Cameroon after cassava diet. Ann. Endocrinol. 38, 80A.

Astwood, E.B. 1943. The chemical nature of compounds which inhibit the function of the thyroid gland. J. Pharmacol. Exp. Ther. 78, 79–89.

Auriga, M., and Koj, A. 1975. Protective effect of rhodanese on the respiration of isolated mitochondria intoxicated with cyanide. Bull. Acad. Pol. Sci. (Biol.), C1. II Vol. XXIII, 5, 305–310.

Azukizawa, M., Murata, Y., Ikenoue, T., Martin Jr., C.B., and Hershman, M. 1976. Effect of thyrotropin-releasing hormone on secretion of thyroptropin, prolactin, thyroxine and triiodothyronine in pregnant and fetal Rhesus monkeys. J. Clin. Endocrinol. Metab. 43, 1020–1028.

Barker, M.H. 1936. The blood cyanates in the treatment of hypertension. J.A.M.A. 106, 762–767.

Barker, S.B., Humphrey, M.J., and Soley, M.H. 1951. Clinical determination of protein-bound iodine. J. Clin. Invest. 30, 55–62.

Barnaby, C.F., Davidson, A.M., and Plaskett, L.G. 1965. Intrathyroidal iodine metabolism in the rat. The influence of diet and administration of thyroid-stimulating hormone. Biochem. J. 95, 811–818.

Barrett, M.D.P. 1976. Dietary cyanide, linamarin and nutritional deficiencies. Ph.D. Thesis, University of Guelph.

Barrett, M.P.D., Alexander, J.C., and Hill, D.C. 1978. Utilization of ^{35}S from radioactive methionine or sulfate in the detoxification of cyanide by rats. Nutr. Metab. 22, 51–57.

Barrett, M.D., Hill, D.C., Alexander, J.C., and Zitnak, A. 1977. Fate of orally dosed linamarin in the rat. Can. J. Physiol. Pharmacol. 55, 134–136.

Belshaw, B.E., and Becker, D.V. 1973. Necrosis of follicular cells and discharge of thyroidal iodine induced by administering iodide to iodine-deficient rats. Endocrinology 36, 466–474.

Ben Yehoshua, S., and Conn, E.E. 1964. Biosynthesis of prunasin, the cyanogenic glucoside of peach. Plant. Physiol. 39, 331–333.

Bervoets, W., and Lassange, M. 1959. Modes et coutumes alimentaires des Congolais en milieu rural. Académie Royale des Sciences Coloniales, Bruxelles. Mémoires, Nouvelle Série, IX, fasc. 4, 1–100.

Bissett, F.H., Clapp, R.C., Coburn, R.A., Ettlinger, M.G., and Long Jr., L. 1969. Cyanogenesis in manioc: concerning lotaustralin. Phytochemistry 8, 2235–2247.
Bode, H.H., Vanjonack, W.J., and Crawford, J.D. 1978. Mitigation of cretinism by breast-feeding. Pediatrics 62, 13–16.
Bolhuis, G.G. 1954. The toxicity of cassava roots. Neth. J. Agric. Sci. 2, 176–185.
Book, S.A., Wolf, H.G., Parker, H.R., and Bustad, L.K. 1974. The exchange of radioiodine in pregnant and fetal sheep. Health Phys. 26, 533–539.
Boulos, B.M., Hanna, F., Davis, L.E., and Almond, C.H. 1973. Placental transfer of antipyrine and thiocyanate and their use in determining maternal and fetal body fluids in a maintained pregnancy. Arch. Int. Pharmacodyn. Ther. 201, 42–51.
Bourdoux, P., Delange, F., Gérard, M., Mafuta, M., Hanson, A., and Ermans, A.M. 1978. Evidence that cassava ingestion increases thiocyanate formation: a possible etiologic factor in endemic goiter. J. Clin. Endocrinol. Metab. 46, 613–621.
Boute, J. 1973. Zaire, pp 803–826. In: Croissance démographique et évolution socio-économique en Afrique de l'Ouest. Recueil d'articles inédits publié sous la direction de John C. Caldwell, avec la collaboration de N.O. Addo, A. Igun et P.O. Olusanya. The Population Council, N.Y., 1973, 1–1028.
Boxer, G.E. and Rickards, J.C. 1952. Studies on the metabolism of the carbon of cyanide and thiocyanate. Arch. Biochem. Biophys. 39, 7–26.
Brown-Grant, K. 1956. Inhibition of iodine-concentrating mechanism in mammary gland by thiocyanate and other anions. Lancet 2, 497.
1957. The iodide concentrating mechanism of the mammary gland. J. Physiol. 135, 644–654.
1961. Extrathyroidal iodide concentrating mechanisms. Physiol. Rev. 41, 189–213.
Brown-Grant, K., and Galton, V.A. 1958. Iodinated compounds in milk after radioiodide administration. Biochim. Biophys. Acta 27, 422–423.
Brownlee, B.E.W., Marchant, B., and Alexander, W.D. 1971. Placental transfer of ^{35}S-propylthiouracil and ^{35}S-methinazole in the rat. In: Further Advances in Thyroid Research. Fellinger, K., and Höfer, R., ed., Wiener Medizinische Akademie Publ., Vienna, 143-147.
Buist, N.R., Murphy, W.F., Brandon, G.R., Foley, T.P., and Penn, R.L. 1975. Neonatal screening for hypothyroidism. Lancet 2, 872–873.
Burssens, H. 1958. Les peuplades de l'Entre Congo-Ubangi (Ngbandi, Ngbaka, Mbanda, Ngombe et Gens d'Eau). Annales du Musée Royal du Congo Belge, Tervuren, Belgique, 1958. Monographies ethnographiques, 1–219.
Buthieau, A.M. and Autissier, N. 1977. Action des ions Mn^{2+} sur le métabolisme iodé thyroïdien. C.R. Soc. Biol. (Paris) 171, 1024–1028.
Butler, G.W. 1965. The distribution of the cyanoglucosides linamarin and lotaustralin in higher plants. Phytochemistry 4, 127–131.
Butler, G.W., Bailey, R.W., and Kennedy, L.D. 1965. Studies on the glucosidase "linamarase". Phytochemistry 4, 369–381.
Buttfield, I.H., and Hetzel, B.S. 1967. Endemic goiter in eastern New Guinea. With special reference to the use of iodized oil in prophylaxis and treatment. Bull. WHO 36, 243–262.
1969. Endemic goiter in New Guinea and the prophylactic program with iodinated poppyseed oil. In: Endemic goiter. Stanbury, J.B. ed., Pan American Health Organization, Washington, Scientific Publication 193, 132–144.
Butts, W.C., Kuehneman, M., and Middowson, G.M. 1974. Automated method for determining serum thiocyanate, to distinguish smokers from nonsmokers. Clin. Chem. 20, 1344–1348.
Buzina, R. 1970. Ten years of goiter prevention in Croatia, Yugoslavia. Am. J. Clin. Nutr. 23, 1085–1089.
Care, A.D. 1954. Goitrogenic activity in linseed. N.Z. J. Sci. Technol. 36, 321–327.
Carr, E.A., Beierwaltes, W.H., Govind Raman, M.B.S., Dodson, V., Tanton, J., Betts, J.S., and Stambaugh, R.A. 1959. The effect of maternal thyroid function on fetal thyroid function and development. J. Clin. Endocrinol. Metab. 19, 1–18.
Cerighelli, R. 1955. Plantes vivrières. In: Cul-

tures Tropicales. Librairie Baillière et Fils, Paris, 289–378.

Chesney, A.M., Clawson, T.A., and Webster, B. 1928. Endemic goiter in rabbits. I. Incidence and characteristics. Bull. Johns Hopkins Hosp. 43, 261–277.

Chopra, I.J., Hershman, J.M., and Hornabrook, R.W. 1975. Serum thyroid hormone and thyrotropin levels in subjects from endemic goiter regions of New Guinea. J. Clin. Endocrinol. 40. 326–333.

Choufoer, J.C., van Rhijn, M., Kassenaar, A.A.H., and Querido, A. 1963. Endemic goiter in Western New Guinea: iodine metabolism in goitrous and non-goitrous subjects. J. Clin. Endocrinol. Metab. 23, 1203–1217.

Clapp, R.C., Bissett, F.H., Coburn, R.A., and Long Jr., L. 1966. Cyanogenesis in manioc: linamarin and isolinamarin. Phytochemistry 5, 1323–1326.

Clements, F.W., and Wishart, J.W. 1956. A thyroid-blocking agent in the etiology of endemic goiter. Metabolism 5, 623-639.

Colle. 1923. Les Gombes de l'Équateur. Histoire et migrations. Bulletin de la Société Royale Belge de Géographie, Bruxelles, XLVII, 1, 141–169.

Conn, E.E. 1973a. Biosynthesis of cyanogenic glycosides. Biochem. Soc. Symp. 38, 277–302.

1973b. Cyanogenic glycosides: their occurrence, biosynthesis and function. In: Chronic Cassava Toxicity. Nestel, B., and MacIntyre, R., ed., International Development Research Centre, Ottawa, IDRC–010e, 55–63.

1969. Cyanogenic glycosides. Agr. Food Chem. 17, 519–526.

1978. Cyanogenesis, the production of hydrogen cyanide, by plants. In: Effects of poisonous plants on livestock. Keeler, R.F., Van Kampen, K.R., and James, L.F., ed., Academic Press, N.Y., 86–95.

Conn, E.E., and Akazawa, T. 1958. Biosynthesis of p-hydroxybenzaldehyde. Fedn. Proc. Fedn. Am. Soc. Exp. Biol. 17, 205.

Conn, E.E., and Butler, G.W. 1969. The biosynthesis of cyanogenic glucosides and other simple nitrogen compounds. In: Perspectives in phytochemistry. Harborne, J.B., and Swain, T., ed., Academic Press, London, Chap. 2, 47–74.

Connolly, R.J. 1973. The changing age incidence of iodbasedow in Tasmania. Med. J. Aust. 2, 141–174.

Connolly, R.J., Vidor, G.I., and Stewart, J.C. 1970. Increase in thyrotoxicosis in endemic goiter area after iodation of bread. Lancet 1, 500–502.

Cooke, R.D., Blake, G.G., and Battershill, J.M. 1978. Purification of cassava linamarase. Phytochemistry 17, 381–383.

Coop, I.E., and Blakley, R.L. 1949. The metabolism and toxicity of cyanides and cyanogenetic glucosides in sheep. I. Activity in the rumen. N.Z. J. Sci. Technol. 30A, 277–291.

Cooper-Driver, G.A., and Swain, T. 1976. Cyanogenic polymorphism in bracken in relation to herbivore predation. Nature 260, 604.

Cornil, J., Ledent, G., Vanderstappen, R., Herman, P., Vandervelden, M., and Delange, F. 1974. Étude comparative de la composition chimique de végétaux et de sols des régions goitreuse et non goitreuse de l'île Idjwi (Lac Kivu, République du Zaïre). Bulletin des Séances de l'Académie Royale des Sciences d'Outre-Mer 3, 386–402.

Costa, A., Cottino, F., Malvano, R., Magro, G., Zoppetti, G., and Buzzigoli, G. 1969. Research into the iodine content of extrathyroidal tissues in man. Folia Endocrinol. 22, 486–497.

Costa, A., Ferro Luzzi, G.F., Marocco, F., Cottino, F., Gianti, S., Patrito, G., Zoppetti, G., Magro, G., Buccini, G., and Balsamo, A. 1967. An investigation of endemic goitre in some piedmontese valleys. Panminerva Med. 9, 55–62.

Coursey, D.G. 1973. Cassava as food: toxicity and technology. In: Chronic Cassava Toxicity. Nestel, B., and MacIntyre, R., ed., International Development Research Centre, Ottawa, IDRC–10e, 1973, 27–36.

Crawley, F.E.H., and Goddard, E.A. 1977. Internal dose from carbon-14 labelled compounds. The metabolism of carbon-14 labelled potassium cyanide in the rat. Health Phys. 32, 135–142.

Croxson, M.S., Gluckman, P.D., and Ibbertson, H.K. 1976. The acute thyroidal response to iodized oil in severe endemic

goitre. J. Clin. Endocrinol. Metab. 42, 926–930.
Culliton, B.J. 1973. Sloan-Kettering: the trials of an apricot pit-1973. Science 182, 1000-1003.
D'Angelo, S.A., and Wall, N.R. 1971. Simultaneous effects of thyroid and adrenal inhibitors on maternal-fetal endocrine interrelations in the rat. Endocrinology 89, 591-597.
Dasen, P.R. 1974. Le développement psychologique du jeune enfant africain. Archives de psychologie 164, 341-386.
Davenport, J.W. 1970. Cretinism in rats: enduring behavioral deficit induced by tricyanoaminopropene. Science, 167, 1007–1009.
De Bruijn, G.H. 1971. Étude du caractère cyanogénétique du manioc (*Manihot esculenta* Crantz). Mededelingen Landbouwhogeschool — Wageningen — Nederland 1971, 1–140.
De Cocker, M. 1950. Essai de parallélisme biblico-congolais. Zaïre. Bruxelles 1950, 3, 277–298.
De Groot, L.J., Jaksina, S., and Karmakar, M., 1968. Role of pyridoxal phosphate in thyroid hormone synthesis. Endocrinology 83, 1253–1258.
Delange, F. 1966. Le goitre endémique de l'île d'Idjwi (Lac Kivu, République du Congo). Données préliminaires. Ann. Endocrinol. (Paris) 27, 256–261.
Delange, F. 1974. Endemic goitre and thyroid function in Central Africa. Monographs in Pediatrics 2, S. Karger, Basel, 1–171.
Delange, F., Beckers, C., Hofer, R., Konig, M.P., Monaco, F., and Varrone, S. 1979. Neonatal screening for congenital hypothyroidism in Europe. Report of the newborn committee of the European Thyroid Association. Acta Endocrinol. 90, suppl. 223, 1–29.
Delange, F., Camus, M., and Ermans, A.M. 1972a. Circulating thyroid hormones in endemic goitre. J. Clin. Endocr. 34, 891–895.
Delange, F., Costa, A., Ermans, A.M., Ibbertson, H.K., Querido, A., and Stanbury, J.B. 1972b. A survey of the clinical and metabolic patterns of endemic cretinism. In: Human Development and the Thyroid Gland. Stanbury, J.B., Kroc, R.L., eds., Plenum Press, New York, 175–187.
Delange, F., Dodion, J., Wolter, R., Bourdoux, P., Dahlem, A., Glinoer, D., and Ermans, A.M. 1978. Transient hypothyroidism in the newborn infant. J. Pediatr. 92, 974–976.
Delange, F., and Ermans, A.M. 1971a. Role of a dietary goitrogen in the etiology of endemic goiter on Idjwi Island. Am. J. Clin. Nutr. 24, 1354–1360.
1971b. Further studies on endemic cretinism in Central Africa. Horm. and Metab. Research 3, 431–436.
1976. Endemic goitre and cretinism. Naturally occurring goitrogens. Pharmac. Ther. C. 1, 57–93.
Delange, F., Ermans, A.M., Vis, H.L., and Stanbury, J.B. 1972c. Endemic cretinism in Idjwi Island, (Kivu Lake, Republic of the Congo). J. Clin. Endocrinol. Metab. 34, 1059–1066.
Delange, F., Hershman, J.M., and Ermans, A.M. 1971. Relationship between the serum thyrotropin level, the prevalence of goiter and the pattern of iodine metabolism in Idjwi Island. J. Clin. Endrocinol. Metab. 33, 261–268.
Delange, F., Thilly, C., Camus, M., Berquist, H., Cremer, N., Hesch, R.D., and Ermans, A.M., 1976. Evidence for foetal hypothyroidism in severe endemic goitre. In: Thyroid Research. Robbins, J. and Braverman, L.E., ed., Excerpta Medica, Amsterdam, 493–496.
Delange, F., Thilly, C., and Ermans, A.M. 1968. Iodine deficiency, a permissive condition in the development of endemic goitre. J. Clin. Endocrinol. Metab. 28, 114–116.
Delange, F., Thilly, C., Pourbaix, P., and Ermans, A.M. 1969. Treatment of Idjwi Island endemic goitre by iodized oil. In: Endemic Goitre. Stanbury, J.B., ed., PAHO, Washington, Scientific Publication 193, N.Y., 118–131.
Delange, F., Van der Velden, M., and Ermans, A.M. 1973. Evidence of an antithyroid action of cassava in man and in animals. In: Chronic Cassava Toxicity. Nestel, B., and MacIntyre, R., ed., International Development Research Centre, Ottawa, IDRC–10e, 147–151.
DeLeon, R., and Rettana, O.G. 1974. Eradication of endemic goiter as a public health

problem in Guatemala. In: Endemic Goiter and Cretinism: Continuing Threats to World Health. Dunn, J.T. and Medeiros-Neto, G.A., ed., PAHO, Washington, Scientific Publication 292, 227–230.

De Smet, M.P. 1960. Contribution à l'étude de l'endémie goitreuse des Uélés (Republique du Congo). Ann. Soc. Belge Med. Trop. 40, 601–622.

De Visscher, M., Beckers, M., Van Den Schrieck, M.G., De Smet, M., Ermans, A.M., Galperin, H., and Bastenie, P.A. 1961. Endemic goitre in the Uele region (Republic of Congo) I. General aspects and functional studies. J. Clin. Endocrinol. Metab. 21, 175–188.

Dilleman, G. 1958. Composés cyanogénétiques. In: Handbuch der Pflanzenphysiologie, Vol. VIII. Springer, Berlin, 1050–1074.

Dodge, P., Palkers, H., Fierro-Benitez, and Ramirez, I. 1969. Effect on intelligence of iodine in oil administered to young Andean children, a preliminary report. In: Endemic Goitre. Stanbury, J.B., ed., PAHO, Washington, Scientific Publication 193, N.Y., 378–380.

Drillien, C.M. 1972. Abnormal neurologic signs in the first year of life in low-birth weight infants: possible prognostic significance. Dev. Med. Child Neurol. 14, 575–584.

Dumont, J.E., Ermans, A.M., and Bastenie, P.A. 1963a. Thyroidal function in a goiter endemic. IV. Hypothyroidism and endemic cretinism. J. Clin. Endocrinol. Metab. 23, 325–335.

1963b. Thryoid function in a goiter endemic. V. Mechanism of thyroid failure in the Uele endemic cretins. J. Clin. Endocrinol. Metab. 23, 847–860.

Dunstan, W.R., and Henry, T.A. 1903. Cyanogenesis in plants. III. On phaseolunatin, the cyanogenetic glucoside of *Phaseolus lunatus*. Proc. R. Soc. Lond. Biol. Sci. 72, 285–294.

Dunstan, W.R., Henry, T.A., and Auld, S.J.M. 1906. Cyanogenesis in plants. V. The occurrence of phaseolunatin in cassava (*Manihot aipi* and *Manihot utilissima*). Proc. R. Soc. Lond. Biol. Sci. 78, 152–158.

Dussault, J.H., Coulombe, P., Laberge, C., Letarte, J., Guyda, H., and Khoury, K. 1975. Preliminary report on a mass screening program for neonatal hypothyroidism. J. Pediatr. 86, 670–674.

Dussault, J.H., Glorieux, J., Letarte, J., Guyda, H., and Laberge, C. 1978. Preliminary report on psychological development at age one of treated hypothyroid infants detected by the Quebec screening network for metabolic diseases. Clin. Res. 26, suppl. 169A.

Echt, C.R., and Doss, J.F. 1963. Myxedema in pregnancy. Obstet. Gynecol. 22, 615–620.

Eder, H. 1951. Determination of thiocyanate space. Methods Med. Res. 4, 48–53.

Ekpechi, O.L. 1967. Pathogenesis of endemic goitre in Eastern Nigeria. Br. J. Nutr. 21, 537–545.

1973. Endemic goitre and high cassava diets in Eastern Nigeria. In: Chronic Cassava Toxicity. Nestel, B. and MacIntyre, R., ed., IDRC, Ottawa, IDRC–010e, 139–145.

Ekpechi, O.L., Dimitriadou, A., and Fraser, R. 1966. Goitrogenic activity of cassava (a staple Nigerian food). Nature (London) 210, 1137–1138.

Ekpechi, O.L., and van Middlesworth, L. 1973. Iodinated compounds in the thyroids of the offspring of rats maintained on low-iodine diet. Endocrinology 92, 1376–1381.

Epoma, F. 1949. Moeurs et habitudes de chez nous. La Voix du Congolais, Léopoldville, V, 40, 275–278.

Ermans, A.M. 1969. Intrathyroidal iodine metabolism in goiter. In: Endemic Goiter. Stanbury, J.B., ed., PAHO, Washington, Scientific Publication 193, 1–13.

Ermans, A.M., Delange, F., Van der Velden, M., and Kinthaert, J. 1972. Possible role of cyanide and thiocyanate in the etiology of endemic cretinism. In: Human Development and the Thyroid Gland. Stanbury, J.B., and Kroc, R.L., eds., Plenum Press, N.Y., 455–486.

Ermans, A.M., Dumont, J.E., and Bastenie, P.A. 1963. Thyroid function in a goiter endemic: I. Impairment of hormone synthesis and secretion in the goitrous gland. J. Clin. Endocrinol. Metab. 23, 539–549.

Ermans, A.M., Kinthaert, J., and Camus, M., 1968. Defective intrathyroidal iodine metabolism in non toxic goiter: inadequate iodination of thyroglobulin. J. Clin. Endocrinol. Metab. 28, 1307–1316.

Ermans, A.M., Thilly, C., Vis, H.L., and Delange, F. 1969. Permissive nature of iodine deficiency in the development of endemic goiter. In: Endemic Goiter. Stanbury, J.B., ed., PAHO, Washington, Scientific Publication 193, 101–117.

Ermans, A.M., Van der Velden, M., Kinthaert, J., and Delange, F. 1973. Mechanism of the goitrogenic action of cassava. In: Chronic Cassava Toxicity. Nestel, B., and MacIntyre, R., eds., IDRC, Ottawa, IDRC-010e, 153–157.

Etling, N. 1977. Concentration of thyroglobulin, iodine contents of thryoglobulin and of iodoaminoacids in human neonates thryoid glands. Acta Paediatr. Scand. 66, 97–102.

Eyjolfsson, R. 1970. Recent advances in the chemistry of cyanogenic glycosides. Fortschr. Chem. Org. Naturst. 28, 74–108.

Fielder, H., and Wood, J.L. 1956. Specificity studies on the β-mercapto-pyruvate-cyanide transulfuration system, J. Biol. Chem. 222, 387-397.

Fierro-Benitez, R., Ramirez, I., Estrella, E., Jaramillo, C., Diaz, C., and Urresta, J. 1969. Iodized oil in the prevention of endemic goiter and associated defects in the Andean region of Ecuador. In: Endemic Goiter. Stanbury, J.B., ed., PAHO, Washington, Scientific Publication 193, 306–321.

Fierro-Benitez, R., Ramirez, I., Estrella, E., and Stanbury, J.B. 1974a. The role of iodine in intellectual development in an area of endemic goiter. In: Endemic Goiter and Cretinism: Continuing Threats to World Health. Dunn, J.T. and Medeiros-Neto, G.A., ed., PAHO, Washington, Scientific Publication 292, 135-140.

Fierro-Benitez, R., Ramirez, I., Garces, J., Jaramillo, C., Moncayo, F., and Stanbury, J.B. 1974b. The clinical pattern of cretinism as seen in Highland Ecuador. Am. J. Clin. Nutr. 27, 531–543.

Fierro-Benitez, R., Ramirez, I., and Suarez, J. 1972. Effect of iodine correction early in fetal life on intelligence quotient. A preliminary report. In: Human Development and the Thyroid Gland. Relation to endemic cretinism. Stanbury, J.B., and Kroc, R.L., eds., Plenum Press, New York, 239–247.

Finnemore, H., Cooper, J.M., Stanley, M.B., and Cobcroft, J.H. 1938. The cyanogentic constituents of Australian and other plants. Part VII. J. Soc. Chem. Ind. (London) 57, 162–169.

Finnemore, H., and Gledhill, W.C. 1928. The presence of cyanogenic glucosides in certain species of acacia. Aust. J. Pharm. 9, 174–178.

Fisher, D.A. 1975. Thyroid function in the foetus. In: Perinatal Thyroid Physiology and Disease. Fisher, D.A., and Burrow, G.N., eds., Raven Press, N.Y., 21–32.

1976. Hypothyroidism: paediatric aspects. In: The Thyroid. Werner, S.C., and Ingbar, S.H., eds., Harper and Row, N.Y., 3rd edition, 807–831.

Fisher, D.A., and Dussault, J.H. 1974. Development of the mammalian thyroid gland. In: Handbook of Physiology. Section 7. Endocrinology. Vol. III Thyroid. Greer, M.A., and Solomon, D.H., ed., American Physiological Society, Washington, D.C., 21-38.

Fisher, D.A., Dussault, J., Sack, J., and Chopra, I.J. 1977. Ontogenesis of hypothalamic–pituitary–thyroid function and metabolism in man, sheep and rat. Recent Prog. Horm. Res. 35, 59–116.

Fisher, D.A., Hobel, C.J., Garza, R., and Pierce, A. 1970. Thyroid function in the preterm fetus. Pediatrics 46, 208–215.

Fisher, D.A., Odell, W.D., Hobel, C.J., and Garza, R. 1969. Thyroid function in the term fetus. Pediatrics 44, 526–535.

Freeman, A.G. 1969. Vitamin-B_{12} deficiency and diabetic neuropathy. Lancet 2, 963.

Funderburk, C.F. 1966. Studies on the physiological occurrence and metabolism of thiocyanate. Dissertation presented to the Graduate Council of the University of Tennessee, Memphis, 1–213.

Funderburk, C.F., and van Middlesworth, L. 1967. Effect of lactation and perchlorate on thiocyanate metabolism. Am. J. Physiol. 213, 1371–1377.

1968. Thiocyanate physiologically present in fed and fasted rats. Am. J. Physiol. 215, 147–151.

1971. The effect of thiocyanate concentration on thiocyanate distribution and excretion. Proc. Soc. Exp. Biol. Med. 136, 1249–1252.

Gaitan, E., Meyer, J.D., and Merino, H. 1974. Environmental goitrogens in Colombia. In: Endemic Goiter and Cretinism: Continuing Threats to World Health. PAHO, Washington, Scientific Publication 292, 107–117.

Gardner, R.M., Kirkland, J.L., Ireland, J.S., and Stancel, G.M. 1978. Regulation of the uterine response to estrogen by thyroid hormone. Endocrinology 103, 1164–1172.

Geber, M., and Dean, R.F.A. 1957a. Gesell tests on African children. Pediatrics 20, 1055–1065.

1957b. Psychomotor development in African children: the effects of social class and the need for improved tests. Lancet 1, 1216.

Gibb, M.C., Carbery, J.T., Carter, R.G., and Catalinac, S. 1974. Hydrocyanic acid poisoning of cattle associated with sudax grass. N.Z. Vet. J. 22, 127.

Glorieux, M. 1955. La Haute Ngiri. Sa situation, ses problèmes. Aide médicale aux missions, Bruxelles 1955, XXVII, 4, 131–135.

Goldstein, F., and Rieders, F. 1953. Conversion of thiocyanate to cyanide by erythrocytic enzyme. Am. J. Physiol. 173, 287–290.

Gordon, J.E., and Ingalls, T.H. 1963. Preventive medicine and epidemiology. Am. J. Med. Sci. 246, 354–376.

Goslings, B.M., Djomokoeljanto, R., Docter, R., Van Hardeveld, C., Hennemann, G., Smeenk, D., and Querido, A. 1977. Hypothyroidism in an area of endemic goiter and cretinism in Central Java, Indonesia. J. Clin. Endocrinol. Metab. 44, 481–490.

Goslings, B.M., Djomokoeljanto, R., Hoedijono, R., Soepardjo, H., and Querido, A. 1975. Studies on hearing loss in a community with endemic cretinism in Central Java, Indonesia. Acta Endocrinol. 78, 705–713.

Gray, B., and Galton, V.A. 1974. The transplacental passage of thyroxine and foetal thryoid function in the rat. Acta Endocrinol. 75, 725–733.

Green, W.L. 1978. Mechanisms of action of antithyroid compounds. In: The Thyroid: A Fundamental and Clinical Text. Werner, S.C., and Ingbar, S.H., ed., Harper and Row, N.Y., 77–87.

Greene, L.S. 1973. Physical growth and development, neurological maturation and behavioral functioning in two Andean communities in which goiter is endemic. Am. J. Phys. Anthropol. 38, 119–134.

Greenman, G.W., Gabrielson, M.O., Howard-Flanders, J., and Wessel, M.A. 1962. Thyroid dysfunction in pregnancy. N. Eng. J. Med. 267, 426–431.

Greer, M.A. 1962. The natural occurrence of goitrogenic agents. Recent Prog. Horm. Res. 18, 187–219.

Greer, M.A., Panton, P., and Greer, S.E. 1975. The effect of iodine deficiency on thyroid function in the infant rat. Metabolism 24, 1391–1402.

Greulich, W.W., and Pyle, S.I. 1959. Radiographic atlas of skeletal development of hand and wrist. Stanford University Press, Stanford, CA.

Gross, J. 1962. Iodine and bromine. In: Mineral Metabolism. An Advanced Treatise. Vol. II. The Elements. Colmar, C.L., and Bronniere, F., ed., Academic Press, London, 221–285.

Grosvenor, C.E. 1960. Secretion of I^{131} into milk by lactating mammary glands. Am. J. Physiol. 199, 419–422.

Guilmin, M. 1972. Quelques coutumes matrimoniales des peuplades de l'Entre-Congo Ubangi. Congo, revue générale de la colonie Belge, Bruxelles I, 1, 44–48.

1933. Quelques proverbes des "Bwaka" expliqués par eux-mêmes. Congo, revue générale de la colonie Belge. Bruxelles II, 4, 535–557.

1947. La polygamie sous l'Équateur. Zaïre. Bruxelles I, 9, 1001–1023.

Habermann, J., Heinze, H.G., Horn, K., Kantlehner, R., Marschner, I., Neumann, J., and Scriba, P.C. 1975. Alimentärer Jodmangel in der Bundesrepublik Deutschland. Dtsch. Med. Wochenschr. 100, 1937-1945.

Halmi, N.S. 1961. Thyroidal iodide transport. Vitam. Horm. 19, 133–163.

Halmi, N.S., King, L.T., Winder, R.R., Hass, A.C., and Stuelke, R.J. 1958. Renal excretion of radioiodide in rats. Am. J. Physiol. 193, 379–385.

Halstrom, F., and Moller, K.D. 1945. Content of cyanide in human organs from cases of poisoning with cyanide taken by mouth, with contribution to toxicology of cyanides. Acta Pharmacol. Toxicol. 1, 18–28.

Hegnauer, R. Chemotaxonomie der Pflanzen (vol. 1-6), eine Übersicht über die Verbreitung der Pflanzen-stoffe. Birkhauser, Basel, 1963–1973.

Hehrmann, R., and Schneider, C. 1974. Der Radioimmunoassay für Trijodothyronin und Thyroxin in Serum und seine Anwendung bei Hyperthyreosen. Der Radiologie 14, 156–160.

Hesch, R.D., and Hüfner, M. 1972. Highly specific antibodies to triiodothyronine. Acta Biol. Med. Ger. 28, 861–864.

Hesch, R.D., Hüfner, M., von zur Mühlen, A., and Köbberling, J. 1974. Radioimmunochemical measurement of thyroid hormones and thyroxine-binding globulin. In: Radioimmunoassay and Related Procedures in Medicine. Vol. II, International Atomic Energy Agency, Vienna, 161–176.

Hetzel, B.S. 1974. The epidemiology, pathogenesis and control of endemic goitre and cretinism in New Guinea. N.Z. Med. J. 80, 482–484.

Hill, G.J., Shine, T.E., Hill, H.Z., and Miller, C. 1976. Failure of amygdalin to arrest B16 melanoma and BW5147 AKR leukemia. Cancer Res. 36, 2102–2107.

Himwich, W.A., and Saunders, J.P. 1948. Enzymatic conversion of cyanide to thiocyanate. Am. J. Physiol. 153, 348–354.

Honour, A.J., Myant, N.B., and Rowlands, E.N. 1952. Secretion of radioiodine in digestive juices and milk in man. Clin. Sci. 11, 447–462.

Hüfner, M., and Hesch, R.D. 1973. A comparison of different compounds for TGB-blocking used in radioimmunoassay for triiodothyronine. Clin. Chim. Acta 44, 101–107.

Humbert, J.R., Tress, J.H., and Branco, K.T. 1977. Fatal cyanide poisoning: accidental ingestion of amygdalin. J.A.M.A. 238, 482.

Ibbertson, H.K., Gluckman, P.D., Croxson, M.S., and Stang, L.J.W. 1974. Goiter and cretinism in the Himalayas: a reassessment. In: Endemic Goiter and Cretinism: Continuing Threats to World Health. Dunn, J.T., and Medeiros-Neto, G.A., ed., PAHO, Washington, Scientific Publication 292, 129–134.

Ibbertson, H.K., Pearl, M., McKinnon, J., Tait, J.M., Lim, T., and Gill, M.B. 1971. Endemic cretinism in Nepal. In: Endemic Cretinism. Hetzel, B.S., and Pharoah, P.O.D., ed., Inst. of Human Biology, Mon. Series, Papua New Guinea, 71–88.

Illig, R. 1979. Congenital hypothyroidism. Clin. Endocrinol. Metab. 8, 49–61.

Inoue, K., and Taurog, A. 1968. Acute and chronic effects of iodide on thyroid radioiodine metabolism in iodine-deficient rats. Endocrinology 83, 279–290.

Jagiello, G.M., and McKenzie, J. 1960. Influence of propylthiouracil on the thyroxine-thyrotropin interplay. Endocrinology 67, 451–456.

Jansz, E.R., Jeyaraj, E.E., Pieris, N., and Abeyratne, D.J. 1974a. Cyanide liberation from linamarin. J. Natl. Sci. Coun. Sri Lanka 2, 57–65.

Jansz, E.R., and Nethsingha, C. 1978. Manioc: selected topics. J. Natl. Sci. Coun. Sri Lanka 1, 83–96.

Jansz, E.R., Pieris, N., Jeyaraj, E.E., and Abeyratne, D.J. 1974b. Cyanogenic glucoside content of manioc. II. Detoxification of manioc chips and flour. J. Natl. Sci. Coun. Sri Lanka 2, 129–134.

Jelliffe, D.B. 1969. Appréciation de l'état nutritionnel des populations (principalement par voie d'enquête dans les pays en développement). Série de monographies n° 53, O.M.S., Genève, 1–286.

Joachim, A.W.R., and Pandittesekere, D.G. 1944. Investigation of the hydrocyanic acid content of manioc (*Manihot utilissima*). Trop. Agric. (Colombo), 100, 150–163.

Jones, W.S. 1969. Thyroid function in human pregnancy. Premature deliveries and reproductive failures of pregnant women with low serum butanol-extractable iodines. Am. J. Obstet. Gynecol. 104, 898–909.

Jorissen, A., and Hairs, E. 1891. La linamarine. Nouveau glucoside fournissant de l'acide cyanhydrique par dédoublement et retiré du Linum usitalissinum. Bull. Acad. R. Sci. Belg. Cl. Sci. 21, 529–539.

Kaellis, E. 1970. Effect of manganous ions on

thyroidal iodine metabolism in the rat. Proc. Soc. Exp. Biol. Med. 135, 216–218.

Kelly, F.C., and Snedden, W.W. 1962. Fréquence et répartition géographique du goitre endémique. In: Le goitre endémique. Série des monographies, O.M.S., Genève, 27–241.

Ketelbant-Balasse, C., Branders, C., and Delange, F. 1978. Morphological and functional study of the thyroid gland in the newborn (Abstract). Ann. Endrocinol. 39, 42A.

Kevany, J., and Chopra, J.G. 1970. The use of iodized oil in goiter prevention. Am. J. Public Health 60, 919–925.

Klein, A.H., Meltzer, S., and Kenny, F.M. 1972. Improved prognosis in congenital hypothyroidism treated before age three months. J. Pediatr. 81, 912–915.

Klein, A.H., Agustin, A.V., and Foley, T.P. 1974. Successful laboratory screening for congenital hypothyroidism. Lancet 2, 77–79.

Knowles, J.A. 1965. Excretion of drugs in milk. A review. J. Pediatr. 66, 1068–1082.

Koutras, D.A. 1974. Variation in incidence of goiter within iodine deficient populations. In: Endemic Goiter and Cretinism: Continuing Threats to World Health. PAHO, Washington, Scientific Publication 292, 95–101.

Kreutler, P.A., Varbanov, V., Goodman, W., Olaya, G., and Stanbury, J.B. 1978. Interactions of protein deficiency, cyanide, and thiocyanate on thyroid function in neonatal and adult rats. Am. J. Clin. Nutr. 31, 282–289.

Lambrechts, A., and Bernier, G. 1961. Enquête alimentaire et agricole dans les populations rurales du Haut Katanga (1957–1958). Fulreac, Liège, 1–236.

Lang, K. 1933. Die Rhodanbildung in Tierkörpor. Biochem. Z. 259, 243–256.

Langer, P. 1964. Serum thiocyanate level in large sections of the population as an index of the presence of naturally occurring goitrogens in the organism. In: Naturally Occurring Goitrogens and Thyroid Function. Podoba, J., and Langer, P., ed., Publishing House of the Slovak Academy of Sciences, Bratislava, 281–295.

Langer, P., Kokesova, H., and Gschwendtova, K., 1976. Acute redistribution of thyroxine after the administration of univalent anions, salicylate, theophylline and barbiturates in rat. Acta Endocrinol. 81, 516–524.

Langer, P., and Kutka, M. 1964. Influence of cabbage on the thyroid function in man (preliminary communication). In: Naturally Occurring Goitrogens and Thyroid Function. Podoba, J., and Langer P., ed., Publishing House of the Slovak Academy of Sciences, Bratislava, 303–306.

Langer, P., and Michajlovskij, N. 1972. Effect of naturally occurring goitrogens on thyroid peroxidase and influence of some compounds on iodide formation during the estimation. Endocrinol. Exp. 6, 97–103.

Leontovitch, C. 1933. L'agriculture indigène dans l'Ubangi. Bulletin agricole du Congo Belge, Bruxelles, XXIV, 1, 45–68.

Lewis, J.P. 1977. Laetrile. West. Med. 127, 55–62.

Ling, E.R., Kon, S.K., and Porter, J.W.G. 1961. The composition of milk and the nutritive value of its components. In: The Mammary Gland and its Secretion. Kon, S.K., and Kowie, A.T., ed., Academic Press, N.Y., Vol. 2, 196–263.

Logothetopoulos, J., and Scott, R.F. 1955. Concentration of iodine 131 across the placenta of the guinea pig and the rabbit. Nature 175, 775–776.

Logothetopoulos, J.H., and Myant, N.B. 1956a. Concentration of radioiodide and ^{35}S-labelled thiocyanate by the stomach of the hamster. J. Physiol. 133, 213–219.

1956b. Concentration of radioiodide and ^{35}S-labelled thiocyanate by the salivary glands. J. Physiol. 134, 189–194.

London, W.T., Money, W.L., and Rawson, R.W. 1964. Placental transfer of ^{131}I-labelled iodide in the guinea pig. J. Clin. Endocrinol. Metab. 28, 247–252.

Lorscheider, F.L., and Reineke, E.P. 1972. Thyroid hormone secretion rate in the lactating rat. J. Reprod. Fertil. 30, 269–279.

Lubchenco, L.A., Horner, F.A., Reed, L.H., Hix, I.E., Metcalf, D., Cohig, R., Eliott, H.C., and Bourg, M. 1963. Sequelae of premature birth. Am. J. Dis. Child. 106, 101–115.

Macy, I.G., and Kelly, H.J. 1961. Human milk and cow's milk in infant nutrition. In: The Mammary Gland and its Secretion. Kon,

S.K., and Kowie, A.T., ed., Academic Press, N.Y., Vol. 2, 265–304.

Maenpaa, J. 1972. Congenital hypothyroidism: aetiological and clinical aspects. Arch. Dis. Child. 47, 914–923.

Maesen, C. 1949. Les Banza. Monographie ethnographique. Status quaestionis. Documentation non publiée déposée au Musée de Tervuren (Service de documentation ethnographique).

Makene, W.J., and Wilson, J. 1972. Biochemical studies in Tanzanian patients with tropical ataxic neuropathy. J. Neurol. Neurosurg. Psychiatry 35, 31–33.

Malamos, B., Koutras, D.A., Mantzos, J., Chiotak, L., Sfontouris, J., Papadopoulos, S.N., Rigopoulos, G.A., Pharmakiotis, A.D., and Vlassis, G. 1970. Endemic goitre in Greece: effects of iodized oil injection. Metabolism 19, 569–580.

Maloof, F., and Soodak, M. 1959. The inhibition of the metabolism of thiocyanate in the thyroid of the rat. Endocrinology 65, 106–113.

1964. The oxidation of thiocyanate by a cytoplasmic particulate fraction of thyroid tissue. J. Biol. Chem. 239, 1995–2001.

Man, E., Reid, W., Hellegers, A., and Jones, W. 1969. Thyroid function in human pregnancy. Am. J. Obstet. Gynecol. 103, 328–347.

Man, E.B., and Benotti, J. 1969. Butanol-extractable iodine in human and bovine colostrum and milk. Clin. Chem. 15, 1141–1146.

Matovinovic, J., Child, M.A., Nichaman, N.Z., and Trowbridge, F.L. 1974. Iodine and endemic goitre. In: Endemic Goiter and Cretinism: Continuing Threats to World Health. Dunn, J.T., and Medeiros-Neto, G.A., ed. PAHO, Washington, Scientific Publication 292, 67–94.

McCarrison, R., and Moldhava, K.B. 1970. Indian Med. Res. Mem. 23, 1923 quoted by E. Kaelis in Proc. Soc. Exp. Biol. Med. 135, 216–218.

McCullagh, S.F. 1963. The huon peninsula endemic: the effectiveness of an intramuscular depot of iodized oil in the control of endemic goitre. Med. J. Aust. 25, 769–777.

Meister, A. 1953. Preparation and enzymic reactions of the keto analozines of asparagine and glutamine. Fed. Proc. 12, 245.

Meister, A., Fraser, P.E., and Tice, S.V. 1954. Enzymatic desulfuration of β-mercaptopyruvate to pyruvate. J. Biol. Chem. 206, 561–575.

Mentzer, C., Favre Bonvin, J., and Massias, M. 1963. Biogenèse du glucoside cyanogénétique chez Prunus laurocerasus. Bull. Soc. Chim. Biol. 45, 749–760.

Michajlovskij, N. 1964. Nahrungsmittel als Rhodanidträger. In: Naturally Occurring Goitrogens and Thyroid Function. Podoba, J., and Langer, P., ed., Publishing House of the Slovak Academy of Sciences, Bratislava, 39–48.

Michajlovskij, N., and Langer, P. 1958. Studien über Beziehungen zwischen Rhodanbildung und kropfbildender Eigenschaft von Nahrungsmitteln. I: Gehalt einiger Nahrungsmittel an präformierten Rhodanid. Z. Physiol. Chem. 312, 26-30.

Miller, J.K., Moss, B.R., Swanson, E.W., and Lyke, W.A. 1975. Effect of thyroid status and thiocyanate on absorption and excretion of iodine by cattle. J. Dairy Sci. 58, 526–531.

Mitchell, M.L., and O'Rourke, M.E. 1960. Response of the thyroid gland to thiocyanate and thyrotropin. J. Clin. Endocrinol. Metab. 20, 47–56.

Mitsuma, T., Colucci, J., Shenkman, L., and Hollander, C.S. 1972. Rapid simultaneous radioimmunoassay for triiodothyronine and thyroxine in unextracted serum. Biochem. Biophys. Res. Commun. 46, 2107–2113.

Monekosso, G.L., and Wilson, J. 1966. Plasma thiocyanate and vitamin B_{12} in Nigerian patients with degenerative neurological disease. Lancet 1, 1062–1064.

Montgomery, R.D. 1964. Observations on the cyanide content and toxicity of tropical pulses. W. Indian Med. J. 13, 1–11.

1965. The medical significance of cyanogen in plant Foodstuffs. Am. J. Clin. Nutr. 17 103–113.

1969. Cyanogens. In: Toxic Constituents of Plant Foodstuffs. Liener, I.E., ed. Academic Press, N.Y., 143–157.

Mora, J.O., Pardo, F., and Rubda-Williamson, R. 1974. Surveillance of salt iodization in Colombia. In: Endemic Goiter and Cretinism: Continuing Threats to World

Health. Dunn, J.T., and Medeiros-Neto, G.A., ed., PAHO, Washington, Scientific Publication 292, 209–213.

Nagataki, S. 1974. Effect of excess quantities of iodide. In: Handbook of Physiology — Section 7 Endocrinology — Volume III. Thyroid. Greer, M.A., and Solomon, D.H., ed., American Physiology Society, Washington D.C., 329–344.

Nartey, F. 1968. Studies on cassava, *Manihot utilissima* Pohl. I. Cyanogenesis: the biosynthesis of linamarin and lotaustralin in etiolated seedlings. Phytochemistry 7, 1307–1312.

Nataf, B., Sfez, M., Michel, R., and Roche, J. 1956. Métabolisme des iodures chez les rattes gestantes et les foetus. Concentration de l'iode radioactif par le placenta. C.R. Soc. Biol. 150, 324–327.

Nestel, B. 1973. Current utilization and future potential for cassava. In: Chronic Cassava Toxicity. Nestel, B., and MacIntyre, R., ed., International Development Research Centre, Ottawa, IDRC-010e, 11–26.

Newell, K.W. 1975. Participation et santé. In: Participation et santé. O.M.S., Genève, 207–219.

Niçaise, J. 1949. Les Ngbandi. Monographie ethnographique. Status quaestionis. Documentation non publiée déposée au Musée de Tervuren.

Nicholls, L. 1951. In: Tropical Nutrition and Dietetics. Baillière, London.

Nuttall, F.O., and Doe, R.P. 1964. The achilles reflex in thyroid disorders. A critical evaluation. Ann. Intern. Med. 61, 269–288.

Nwokolo, C., Ekpechi, O.L., and Nwokolo, U. 1966. New foci of endemic goitre in Eastern Nigeria. Trans. R. Soc. Trop. Med. Hyg. 60, 97–108.

Oddie, T.H., Fisher, D.A., McConahey, V.M., and Thompson, C.S. 1970. Iodine intake in the United States: a reassessment. J. Clin. Endocrinol. Metab. 30, 659–665.

Odell, W.D., Wilbur, J.F., and Paul, W.E. 1965. Radioimmunoassay of thyrotropin in human serum. J. Clin. Endocrinol. Metab. 25, 1179–1188.

1965. Radioimmunoassay of human thyrotropin in serum. Metabolism 14, 465–467.

Ohtaki, S., and Rosenberg, I.N. 1971. Prompt stimulation by TSH of thyroid oxidation of thiocyanate. Endocrinology 88, 566–573.

Oluwasanmi, J.O., and Alli, A.F. 1968. Goiters in Western Nigeria. Trop. Geogr. Med. 20, 357–366.

Ombredane, A., Robaye, F., and Plumail, H. 1956. Résultats d'une application répétée du matrix-couleur à une population de noirs congolais. In: Bulletin du centre d'études et de recherches psychotechniques. 5ème année, 6, 129–147.

Osuntokun, B.O. 1970. Cassava diet and cyanide metabolism in Wistar rats. Br. J. Nutr. 24, 797–800.

1971. Epidemiology of tropical nutritional neuropathy in Nigerians. Trans. R. Soc. Trop. Med. Hyg. 65, 454–479.

Osuntokun, B.O., Aladetoyinbo, A., and Adeuja, A.O.G. 1970. Free-cyanide levels in tropical ataxic neuropathy. Lancet 2, 372–373.

Osuntokun, B.O., Durowoju, J.E., McFarlane, H., and Wilson, J. 1968. Plasma amino-acids in the Nigerian nutritional ataxic neuropathy. Br. Med. J. 3, 647–649.

Osuntokun, B.O., and Monekosso, G.L. 1969. Degenerative tropical neuropathy and diet. Br. Med. J. 3, 178–179.

Osuntokun, B.O., Monekosso, G.L., and Wilson, J. 1969. Relationship of a degenerative tropical neuropathy to diet. Report of a field survey. Br. Med. J. 1, 547–550.

Pales, P., and Tassin De Saint-Pereuse, M. 1953. Le goitre endémique en AOF d'après les enquêtes du service de santé en 1948 et 1950. Ed. Mission anthropologique de l'AOF. Direction de la Santé Publique, Dakar.

Pascual-Leone, A.M., Garcia, M.D., and Hervas, F. 1978. Changes in parameters of growth hormone and thyrotropic hormone, and of thyroid function during the early postnatal period in the rat. Rev. Esp. Fisiol. 34, 301–308.

Perez, C., Scrimshaw, N.S., and Munoz, J.A. 1960. Technique to endemic goitre surveys. In: Endemic Goitre. World Health Organization, Geneva, Monograph Series 44, 369–383.

1962. Technique des enquêtes sur le goitre endémique. In: Le goitre endémique.

O.M.S., Série des monographies N° 44, Genève, 383–398.

Pettigrew, A.R., and Fell, G.S. 1972. Simplified colorimetric determination of thiocyanate in biological fluids, and its application to investigation of the toxic amblyopias. Clin. Chem. 18, 996–1000.

Pettigrew, A.R., Logan, R.W., and Willocks, J. 1977. Smoking in pregnancy — effects on birth weight and on cyanide and thiocyanate levels in mother and baby. Br. J. Obstet. Gynecol. 84, 31–34.

Pharoah, P.O.D., Buttfield, I.H., and Hetzel, B.S. 1971. Neurological damage to the fetus resulting from severe iodine deficiency during pregnancy. Lancet 1, 308-310.

1972. The effect of iodine prophylaxis on the incidence of endemic cretinism. Adv. Exp. Med. Biol. 30, 201–221.

Phillips, T.P. 1974. Cassava utilization and potential markets. International Development Research Centre, IDRC–020e, Ottawa. 182 p.

Pieris, N., and Jansz, E.R. 1975. Cyanogenic glucoside content of manioc. III. Fate of bound cyanide on processing and cooking. J. Natl. Sci. Coun. Sri Lanka 3, 41–50.

Pieris, N., Jansz, E.R., and Kandage, R. 1974. Cyanogenic glucoside content of manioc. I. An enzymic method of determination applied to processed manioc. J. Natl. Sci. Coun. Sri Lanka 2, 67–76.

Piironen, E., and Virtanen, A.I. 1963. The effect of thiocyanate in nutrition on the iodine content of cow's milk. Z. Ernaehrungswiss. 3, 140–147.

Potter, G.D., and Chaikoff, I.L. 1956. Identification of radioiodine compounds eliminated in the milk of lactating rats injected with ^{131}I-iodide. Biochim. Biophys. Acta 21, 400–401.

Potter, G.D., Tong, W., and Chaikoff, I.L. 1959. The metabolism of I^{131}-labeled iodine, thyroxine, and triiodothyronine in the mammary gland of the lactating rat. J. Biol. Chem. 234, 350–354.

Pretell, E.A. 1972. The optimal program for prophylaxis of endemic goiter with iodized oil. In: Human Development and the Thyroid Gland: Relation to Endemic Cretinism. Stanbury, J.B., and Kroc, R.L., ed., Plenum Press, N.Y. 267–288.

Pretell, E., Degrossi, O., Riccabona, G., Stanbury, J., and Thilly, C. 1974. The use of iodized oil. In: Endemic Goiter and Cretinism: Continuing Threats to World Health. Dunn, J.T., and Medeiros-Neto, G.A., ed. PAHO, Washington, Scientific Publication 292, 278–281.

Pretell, E.A., Moncloa, F., Salina, R., Guerra-Garcia, R., Kawano, A., Gutierrez, L., Pretell, J., and Wan, M. 1969a. Endemic goiter in rural Peru: effect of iodized oil on prevalence and size of goiter and on thyroid iodine metabolism in known endemic goitrous population. In: Endemic Goiter. Stanbury, J.B. ed. PAHO, Washington, Scientific Publication 193, 419–437.

Pretell, E.A., Moncloa, F., Salina, R., Kawano, A., Guerra-Garcia, R., Gutierrez, L., Beteta, L., Pretell, J., and Wan, M. 1969b. Prophylaxis and treatment of endemic goiter in Peru with iodized oil. J. Clin. Endocrinol. Metab. 29, 1586–1595.

Pretell, E.A., Torres, T., Zenteno, V., and Cornejo, M. 1972. Prophylaxis of endemic goiter with iodized oil in rural Peru. In: Human Development and the Thyroid Gland: Relation to Endemic Cretinism. Stanbury, J.B., and Kroc, R.L., ed., Plenum Press, N.Y., 249–263.

Pyle, S.I., and Hoerr, N.L. 1955. Radiographic atlas of skeletal development of the knee. C.C. Thomas.

Querido, A., Delange, F., Dunn, J.T., Fierro-Benitez, R., Ibbertson, H.K., Koutras, D.A., and Perinetti, H. 1974. Definitions of endemic goiter and cretinism, classification of goiter size and severity of endemias and survey techniques. In: Endemic Goiter and Cretinism: Continuing Threats to World Health. Dunn, J.T., and Medeiros-Neto, G.A., ed., PAHO, Washington. Scientific Publication 292, 267–272.

Quinones, J.D., Boyd, C.M., Beierwaltes, W.H., and Poissant, G.R. 1972. Transplacental transfer and tissue distribution of ^{14}C-2-thiouracil in the fetus. J. Nucl. Med. 13, 148–154.

Raben, M.S. 1949. The paradoxical effect of thiocyanate and of thyrotropin on the organic binding of iodine by the thyroid in the presence of large amounts of iodide. Endocrinology 45, 296–304.

Raiti, S., and Newns, G.H. 1971. Cretinism: early diagnosis and its relation to mental prognosis. Arch. Dis. Child. 46, 692–694.

Rajaguru, A.S.B. 1975. Problem of HCN in cassava. Agricultural Research Seminar Series, 11 December 1975, Paper 75–7; Faculty of Agriculture, University of Sri Lanka Peradeniya, Sri Lanka.

Ramirez, I., Fierro-Benitez, R., Estrella, E., Gomez, A., Jaramillo, C., Hermida, C., and Moncayo, F. 1972. The results of prophylaxis of endemic cretinism with iodized oil in rural Andean Ecuador. In: Human Development and the Thyroid Gland: Relation to Endemic Cretinism. Stanbury, J.B., and Kroc, R.L., ed., Plenum Press, N.Y., 223–235.

Ramirez, I., Fierro-Benitez, R., Estrella, E., Jaramillo, C., Diaz, C., and Urresta, J. 1969. Iodized oil in the prevention of endemic goiter and associated defects in the Andean regions of Ecuador: II. Effects on neuromotor development and somatic growth in children before two years. In: Endemic Goiter. Stanbury, J.B., ed., PAHO, Washington, Scientific Publication 193, 341–359.

Raven, J.C. 1965. Guide to using the coloured progressive matrices. Lewis, H.K. and Co Ltd, London, 1–43.

Redshaw, M.R., and Lynch, S.S. 1974. An improved method for the preparation of iodinated antigens for radioimmunoassay. J. Endocrinol. 60, 527–528.

Reh, E. 1963. Manuel des enquêtes familiales de consommation. FAO. Études de nutrition de la FAO, N° 18.

Reinwein, D. 1961. Die Verteilung der Thiosulfat-schwefeltransferase und des Rhodanids im Menschlichen und tierischen organismus. Z. Physiol. Chem. 326, 94-101.

Riley, M., and Gochman, A. 1964. A fully automated method for the determination of serum protein bound iodine. Technicon Symposium, New York.

Roche, M. 1959. Elevated thyroidal ^{131}I uptake in the absence of goiter in the isolated Venezuelan Indians. J. Clin. Endocrinol. Metab. 19, 1440–1445.

Rodbard, D., Rayford, P.L., Cooper, J.A., and Ross, G.T. 1968. Statistical quality control of radioimmunoassays. J. Clin. Endocrinol. Metab. 28, 1412–1418.

Roger, G., Klees, M., Lagasse, R., Berquist, H., and Thilly, C. 1977. Psychomotor development of children in regions of endemic goiter. Ann. Endocrinol. 38, 50A.

Sack, J., Mado, O.A., and Lunenfeld, B. 1977. Thyroxine concentration in human milk. J. Clin. Endocrinol. Metab. 45, 171–173.

Sadoff, L., Fuchs, K., and Hollander, J. 1978. Rapid death associated with laetrile ingestion. J.A.M.A. 239, 1532.

Sanchez-Martin, J.A., and Mitchell, M.L. 1960. Effect of thyrotropin upon the intrathyroidal metabolism of thiocyanate-S^{35}. Endocrinology 67, 325–331.

Schmidt, M.J., and Allen, J.R. 1967. Effects of 1,1,3-tricyano-2-amino-1 propene on the pre- and postnatal development of the rat. Lab. Invest. 17, 255–264.

Schöberl, A., Kawohl, M., and Hamm, R. 1951. Die Umsetzung von Cystin und Cystamin mit Kaliumcyanid, ein neuer Weg in die Thiazolinchemie. Chem. Ber. 84, 571-576.

Schulz, V., Bonn, R., and Kindler, J. 1979. Kinetics of elimination of thiocyanate in 7 healthy subjects and in 8 subjects with renal failure. Klin. Wochenschr. 57, 243–247.

Schultze, A.B., and Noonan, J. 1970. Thyroxine administration and reproduction in rats. J. Anim. Sci. 30, 774–776.

Scranton, J.R., Nissen, W.M., and Halmi, N.S. 1969. The kinetics of the inhibition of thyroidal iodide accumulation by thiocyanate: a reexamination. Endocrinology 85, 603–607.

Secor, J.B., Conn, E.E., Dunn, J.E., and Seigler, D.S. 1976. Detection and identification of cyanogenic glucosides in six species of acacia. Phytochemistry 15, 1703–1706.

Sedlak, J., Langer, P., Michajlovskij, N., and Kalocai, S. 1964. Correlation between cabbage goitrogenicity and sulphur utilization. In: Naturally Occurring Goitrogens and Thyroid Function." Podoba, J., and Langer, P., ed., Publishing House of the Slovak Academy of Sciences, Bratislava, 161–181.

Seifert, P. 1955. Blausaure-Verbindungen. In: Modern Methods of Plant Analysis. Paech, K., and Tracey, M.V., ed., Vol. IV, Springer, Berlin, 676–688.

Sellers, E.A., and Schönbaum, E. 1962. Goitrogenic action of thyroxine administered with propylthiouracil. Acta Endocrinol. 40, 39–50.

Senecal, J., and Falade, S.A. 1956. Développement psychomoteur de l'enfant africain au cours de la première année. Bull. Med. AOF 1, 300–309.

Silink, K. 1964. Goitrogens in food and endemic goiter. In: Naturally Occurring Goitrogens and Thyroid Function. Podoba, J., and Langer, P., ed., Publishing House of the Slovak Academy of Sciences, Bratislava, 247–269.

Slanina, P., Ullberg, S., and Hammarström, L. 1973. Distribution and placental transfer of ^{14}C-thiourea and ^{14}C-thiouracil in mice studied by whole-body autoradiography. Acta Pharmacol. 32, 358–368.

Smith, D., Blizzard, R.M., and Wilkins, L. 1957. The mental prognosis in hypothyroidism of infancy and childhood. Pediatrics 19, 1011–1022.

Smith, A.M. 1961. Retrobulbular neuritism addisonian pernicious anaemia. Lancet 1, 1001–1002.

Sooch, S.S., Deo, M.G., Karmarkar, M.G., Kochupillai, N., Ramarchandran, K., and Ramalingaswami, V. 1973. Prevention of endemic goitre with iodized salt. Bull. WHO 49, 307–312.

Spindel, E., White, P.K., and Stanbury, J.B. 1976. Roles of fetal and maternal thyroid function in development of young rat. In: Thyroid Research. Robbins, J., and Braverman, L.E., ed., Excerpta Medica, Amsterdam, 505–508.

Srinivasan, V., Mougdal, N.R., and Sarma, P.S. 1957. Studies on goitrogenic agents in foods. I. Goitrogenic action of groundnut. J. Nutr. 61, 87–95.

Städtler, F., and Klemm, W. 1973. Quantitative und histologische Untersuchungen an der fetalen und mütterlichen Schildrüse der ratte unter Propylthiouracil. Beitr. Path. Bd. 149, 293–306.

Stanbury, J.B. 1974. A prospectus for endemic goiter. In: Endemic Goiter and Cretinism: Continuing Threats to World Health. Dunn, J.T., and Medeiros-Neto, G.A., ed., PAHO, Washington, Scientific Publication 292, 291–295.

Stanbury, J.B., Brownell, G.L., Riggs, D.S., Perinetti, H., Itoiz, J., and Del Castillo, E.B. 1954. Endemic goiter. The adaptation of man to iodine deficiency. Harvard University Press, Cambridge, MA, 1–209.

Stanbury, J.B., and Kevany, J.P. 1970. Iodine and thyroid disease in Latin America. Environ. Res. 3, 353–363.

Stanley, M.M., and Astwood, E.B. 1948. The accumulation of radioactive iodide by the thyroid gland in normal and thyrotoxic subjects and the effect of thiocyanate on its discharge. Endocrinology 42, 107–123.

Stewart, J.C., Vidor, G.I., Buttfield, I.H., and Hetzel, B.S. 1971. Endemic thyrotoxicosis in Northern Tasmania: studies of clinical features and iodine nutrition. Aust. N.Z. J. Med. 3, 203–211.

Stolc, V., Knopp, J., and Stolcova, E. 1973. Iodine concentration and content in the organs of rat during postnatal development. Biol. Neonate 23, 35–44.

Strbak, V., Macho, L., Kovac, R., Skultetyova, M., and Michalickova, J. 1976. Thyroxine (by competitive binding analysis) in human and cow milk and in infants' formulas. Endocrinol. Exp. 10, 167–174.

Strbak, V., Macho, L., Uhercik, D., and Kliment, V. 1978. The effect of lactation on thyroid activity of women. Endokrinologie 72, 183–187.

Studer, H., and Greer, M.A. 1968. The regulation of thyroid function in iodine deficiency. Hans Huber Publishers, Bern, 1–119.

Swysen, M. 1978. Analyse bibliographique sur les thiocyanates. Mémoire de Licence en Médecine du Travail, ULB.

Tanghe, B.O. 1929a. De Ngbandi naar het leven geschetst. Congo Bibliotheek, Bruxelles ou Bruges, XXIX.

1929b. De Ngbandi. Geschiedkundige bijdragen. Congo Bibliotheek, Bruxelles ou Bruges, XXX.

Tanghe, B.O. 1930. Le droit d'ainesse chez les indigènes du Haut-Ubangi. Africa, Journal of the International Institute of African Languages and Cultures. London, III, 78–82.

Tauil, M.C., and Azevedo, A.C. 1978. Community participation in health activities in an Amazon community of Brazil. Bull. PAHO 12, 95–103.

Taylor, C.E. 1978. Réorientation du personnel

de santé en fonction des besoins de la population. Carnets de l'enfance, UNICEF, 42, 75–88.

Thilly, C.H., Delange, F., Camus, M., Berquist, H., and Ermans, A.M. 1974. Fetal hypothyroidism in endemic goiter: the probable pathogenic mechanism of endemic cretinism. In: Endemic Goiter and Cretinism: Continuing Threats to World Health. Dunn, J.T., and Medeiros-Neto, G.A., ed., PAHO, Washington, Scientific Publication 292, 121–128.

Thilly, C.H., Delange, F., and Ermans, A.M. 1972. Further investigations of iodine deficiency in the etiology of endemic goiter. Am. J. Clin. Nutr. 25, 30–40.

Thilly, C.H., Delange, F., Lagasse, R., Bourdoux, P., Ramioul, L., Berquist, H., and Ermans, A.M. 1978. Fetal hypothyroidism and maternal thyroid status in severe endemic goiter. J. Clin. Endocrinol. Metab. 47, 354–360.

Thilly, C.H., Delange, F., Golstein-Golaire, J., and Ermans, A.M. 1973a. Endemic goiter prophylaxis by iodized oil: a reassessment. J. Clin. Endocrinol. Metab. 36, 1196–1204.

Thilly, C.H., Delange, F., Ramioul, L., Lagasse, R., Luvivila, K., and Ermans, A.M. 1977. Strategy of goiter and cretinism control in Central Africa. Int. J. Epidemiol. 6, 43–53.

Thilly, C.H., Ramioul, L., and Ermans, A.M. 1973b. A collaborative study on the geographical variations of thyroidal uptake in normal subjects of Europe. Eur. J. Clin. Invest. 3, 272.

Tshibangu, D. 1978. Stratégies et besoins de santé dans la sous-région de l'Ubangi. Mémoire de Licence spéciale en santé publique, Ecole de Santé Publique de l'Université de Bruxelles, 1–132.

Trowbridge, F.L. 1972. Intellectual assessment in primitive societies, with a preliminary report of a study of the effects of early iodine supplementation on intelligence. In: Human Development and the Thyroid Gland: Relation to Endemic Cretinism. Stanbury, J.B., and Kroc, R.L., ed., Plenum Press, N.Y. 137–150.

Vagenakis, A.G., Wang, C.A., Burger, A., Maloof, F., Braverman, L.E., and Ingbar, S.H. 1972. Iodide-induced thyrotoxicosis in Boston. N. Eng. J. Med. 287, 523–527.

Vanderlaan, J.E., and Vanderlaan, W.P. 1947. The iodide concentrating mechanism of the rat thyroid and its inhibition by thiocyanate. Endocrinology 40, 403–416.

Vanderlaan, W.P., and Bissell, A. 1946. Effects of propylthiouracil and of potassium thiocyanate on the uptake of iodine by the thyroid gland of the rat. Endocrinology 39, 157–160.

Van der Velden, M., Kinthaert, J., Orts, S., and Ermans A.M. 1973. A preliminary study on the action of cassava on thyroid iodine metabolism in rats Br. J. Nutr. 30, 511–517.

Van Etten, C.H. 1969. Goitrogens. In: Toxic Constituents of Plant Foodstuffs. Liner, I.E., ed., Academic Press, N.Y., 103–142.

Van Leeuwen, E. 1954. Een vorm van genuine hyperthyreose (Basedow zonder exophtalmus) na gebruik van gejodeerd brood. Ned. Tijdschr. Geneeskd. 98, 81.

Van Middlesworth, L., Jagiello, G., and Vanderlaan, W.P. 1959. Observations on the production of goiter in rats with propylthiouracil and on goiter prevention. Endocrinology 64, 186–190.

Vansina, J. 1966. Introduction à l'ethnographie du Congo. C.R.I.S.P., Bruxelles. Éditions Universitaires du Congo, Kinshasa, 1–228.

Varma, S.K., Collins, M., Row, A., Haller, W.S., and Varma, K. 1978. Thyroxine, triiodothyronine, and reverse triiodothyronine concentrations in human milk. J. Pediatr. 93, 803–806.

Vidor, G.I., Stewart, J.C., Wall, J.R., Wangel, A., and Hetzel, B.S. 1973. Pathogenesis of iodine-induced thyrotoxicosis: studies in Northern Tasmania. J. Clin. Endocrinol. Metab. 37, 901–908.

Vigouroux, E. 1976. Dynamic study of postnatal thyroid function in the rat. Acta Endocrinol. 83, 752–762.

Virtanen, A.I. 1961. Über die Chemie der Brassica-Faktoren, ihre Wirkung auf die Funktion der Schildrüse und ihr Übergehen in die Milch. Experienta (B2501), 17, 241–251.

Virtanen, A.I., and Gmelin, R. 1960. On the transfer of thiocyanate from fodder to milk. Acta Chem. Scand. 14, 941-943.

Vis, H.L., Pourbaix, P., Thilly, C., and van der Borght, H. 1969. Analyse de la situation nutritionnelle de sociétés traditionnelles de la région du lac Kivu: les Shi et les Havu. Enquête de consommation alimentaire. Ann. Soc. Belge Med. Trop. 49, 353–419.

Vis, H.L., Yourassowsky, C., and van der Borght, H. 1972. Une enquête de consommation alimentaire en République Rwandaise. Institut National de Recherche Scientifique. Butare — Rwanda Publication 10, 1–139 + annexes.

Wade, N. 1977. Laetrile at Sloan-Kettering: a question of ambiguity. Science 198, 1231–1234.

Walser, M., and Rahill, W.J. 1965. Nitrate, thiocyanate and perchlorate clearance. Am. J. Physiol. 208, 1158–1164.

Watanabe, T., Moran, D., Tanner, E.E., Staneloni, L., Salvaneschi, J., Altschuler, N., Degrossi, O.J., and Niepominiszcze, H. 1974. Iodized oil in the prophylaxis of endemic goitre in Argentina. In: Endemic Goiter and Cretinism: Continuing Threats to World Health. Dunn, J.T., and Medeiros-Neto, G.A., ed., PAHO, Washington, Scientific Publication 292, 267–280.

Weaver, J.C., Kamm, M.L., and Dobson, R.L. 1960. Excretion of radioiodine in human milk. J.A.M.A. 173, 872–875.

Wells, D.G., Langman, M.J.S., and Wilson, J. 1972. Thiocyanate metabolism in human vitamin B_{12} deficiency. Br. Med. J. 4, 588–590.

Westley, J. 1973. Rhodanese. Adv. Enzymol. Relat. Areas Mol. Biol. 39, 327–368.

Wheeler, J.L., Hedges, D.A., and Till, A.R. 1975. A possible effect of cyanogenic glucoside in sorghum on animal requirements for sulphur. J. Agric. Sci., Camb. 84, 377–379.

Wilkins, L. 1941. Epiphysial dysgenesis associated with hypothyroidism. Am. J. Dis. Child. 61, 13–34.

Wilson, D.C. 1954. Goitre in Ceylon and Nigeria. Br. J. Nutr. 8, 90–99.

Wilson, J. 1965. Leber's hereditary optic atrophy: a possible defect of cyanide metabolism. Clin. Sci. 29, 505–515.

Wilson, J., Linnell, J.C., and Matthews, D.M. 1971. Plasma-cobalamins in neuro-ophthalmological diseases. Lancet 1, 259–261.

Wilson, J., and Matthews, D.M. 1966. Metabolic inter-relationships between cyanide, thiocyanate and vitamin B_{12} in smokers and non-smokers. Clin. Sci. 31, 1–7.

Wokes, F., Baxter, N., Horsford, J., and Preston, B. 1951. Effect of light on vitamin B_{12}. Biochem. J. 53, XIX-XX.

Wolff, J. 1964. Transport of iodide and other anions in the thyroid gland. Physiol. Rev. 44, 45–90.

1976. Iodine homeostasis. In: Regulation of Thyroid Function. Klein, E., and Reinwein, D., ed., F.K. Schattauer Verlag. Stuttgart-New York, 65–78.

Wolff, J., Chaikoff, I.L., Taurog, A., and Rubin, L. 1946. The disturbance in iodine metabolism produced by thiocyanate: the mechanism of its goitrogenic action with radioactive iodine as indicator. Endocrinology 39, 140–148.

Wollman, S.H., and Reed, F.E. 1958. Acute effect of organic binding of iodine on the iodide concentrating mechanism of the thyroid gland. Am. J. Physiol. 194, 28–32.

Wollman, S.H. 1962. Inhibition by thiocyanate of accumulation of radioiodine by thyroid gland. Am. J. Physiol. 203, 517–524.

Wood, J.L. 1975. Biochemistry. In: Chemistry and Biochemistry of Thiocyanic Acid and Its Derivatives. Newman, A.A. ed. Academic Press. London, 156–221.

Wood, J.L., and Cooley, S.L. 1956. Detoxication of cyanide by cystine. J. Biol. Chem. 218, 449–457.

Wood, J.L., and Williams Jr., E.F. 1949. The metabolism of thiocyanate in the rat and its inhibition by PTU. J. Biol. Chem. 177, 59–67.

Wood, T. 1966. The isolation, properties and enzymic breakdown of linamarin from cassava. J. Sci. Food Agric. 17, 85–90.

Yamada, T., Kajimara, A., Takemura, Y., and Onaya, T. 1974. Antithyroid compounds. In: Handbook of Physiology. Section 7. Endocrinology. Vol. III. Thyroid. Greer, M.A., and Solomon, D.H., ed. American Physiological Society, Washington, D.C., 345–357.

Zitnak, A., Hill, D.C., and Alexander, J.C. 1977. Determination of linamarin in biological tissues. Anal. Biochem. 77, 310–314.

Technical editing: Amy Chouinard